CAMS

CAMS
Design, Dynamics, and Accuracy

HAROLD A. ROTHBART

Member of
the Mechanical Engineering Department
The City College of New York
and
Consulting Professional Engineer

NEW YORK · JOHN WILEY & SONS, INC.
LONDON · CHAPMAN & HALL, LIMITED

Library of Congress Catalog Card Number: 56-7163
PRINTED IN THE UNITED STATES OF AMERICA

Dedicated to my mother,

JEANETTE COLT ROTHBART

PREFACE

This book is written primarily for the designer concerned with creating a mathematical function, motion, mechanism, machine, or mechanical computer. The cam mechanism most easily fulfills many of their requirements. The book should also be of value to mechanical engineering students and instructors.

The investigation of this broad subject of cams required both theoretical and practical excursions into: (1) *kinematics*, which considers the relative motions of constrained, rigid members; (2) *dynamics*, which investigates the masses, accelerations, and elasticities of the members; and (3) *machine design*, which resolves the practical construction and proportion of parts and considers the fabrication accuracy.

Besides the diversified cam applications I have aimed the presentation of the advanced chapters toward the expanding field of high-speed machinery. In this category vibrations play an essential part. To date, the subject of cams was regarded only as part of the study of kinematics, and the following relevant points were not included: (1) the shape of the acceleration curve, which provides information on vibrations, noise, and wear; (2) the flexibility and backlash that exist in most systems; and (3) the accuracy of cam profile in terms of its effect on the performance of the mechanism.

The foregoing are pertinent to the study of high-speed operation in which the designer can no longer ignore the necessity of more sophisticated mathematical study, related, of course, to the practical phases of the problems. However, mathematical treatment is kept to a minimum, only that significant to design being presented. "Polydyne Cams" (Chapter 7) and "The Dynamics of High-Speed Cam Systems" (Chapter 8) are of such fundamental nature that the mathematics may be applied to any complex mechanism, cams included. For these chapters the reader should have an understanding of basic vibration theory.

In the later chapters a discussion of profile accuracy and errors is offered. A surface may appear smooth to the eye and yet have poor dynamic properties. In my experience only infrequently does the actual cam agree with the theoretical one. Poor manufacturing techniques can seriously impede the functional ability of a mechanism or machine. No attempt has been made to delve into the various detailed methods of cam fabrication. Accuracy of manufacturing is discussed together with the methods utilized to achieve this precision as determined by experience and the availability of machine tool equipment.

I wish to thank the many in industry who were helpful in providing information for the book. Particular recognition is due the excellent papers by Mr. W. M. Dudley and Mr. D. A. Stoddart on polydyne cams, which were used quite freely in developing and adapting this method. Special appreciation goes to my colleague Mr. Q. Darmstadt, who appraised Chapter 8, Mr. K. Brunell and also Mr. S. Fenster, who reviewed the manuscript and made many valuable suggestions.

HAROLD A. ROTHBART

New York
June, 1956

CONTENTS

NOTATIONS

a	= constant.
a	= $dv/dt = d^2y/dt^2$ = acceleration of follower, in./sec².
a_1, a_2, a_3	= coefficients of respective harmonic number, sin $n\omega t$.
b_1, b_2, b_3	= coefficients of respective harmonic number, cos $n\omega t$.
b	= constant.
b	= part of half rise angle, $\beta/2$.
b_f	= damping factor between linkage and frame lb-sec/in.
b_i	= damping factor in follower linkage, lb-sec/in.
c	= dynamic constant, degree².
c_e	= distance between centers of rotation, in.
d	= distance, in.
d_h	= hub diameter, in.
d_s	= shaft diameter, in.
e	= offset or eccentricity, in.
e	= natural logarithmic base, 2.718.
f	= cam factor for each curve.
f_n	= undamped natural frequency of follower linkage, cycles/sec.
g	= gravitational constant, 32.16 ft/sec².
h	= maximum displacement of follower, in.
j	= constant.
k	= equivalent spring rate of follower linkage, lb/in.
$\bar{k}$	= radius of gyration, in.
k_f	= spring rate of follower linkage, lb/in.
k_r	= equivalent spring rate ratio of follower linkage.
k_s	= spring rate of compression spring, lb/in.
l	= developed length of cam for angle θ, in.
l_a	= length of oscillating follower arm, in.
l_h	= developed length of cam for rise h, in.
m	= equivalent mass at follower end, lb-sec²/in.
n	= any number.

n = frequency ratio.

p = da/dt = pulse of follower, in./sec³.

q = eccentricity of flat-faced follower contact point from cam center, in.

r = radius to trace point, in.

r_a = ramp height, in.

r_c = radial distance to cam profile, in.

r_g = radial distance to center of cutter or grinder, in.

r_k = clearance or backlash in linkage, in.

r_s = initial static deflection of follower linkage, in.

s = distance between cam and oscillating follower centers, in.

t = time for cam to rotate angle θ, sec.

t_h = thickness of contacting cam and follower, in.

t_m = temperature, degree F.

u = distance from centro to cam center of curvature, in.

v = dy/dt = velocity of follower, ips.

v_0 = initial velocity of follower, ips.

v_m = maximum velocity of follower, ips.

w = equivalent weight at follower end, in.

y = displacement of follower, in.

$y' = dy/d\theta$ = follower velocity, in./degree.

$y'' = dy'/d\theta$ = follower acceleration, in./degree².

$y''' = dy''/d\theta$ = follower pulse, in./degree³.

y_c = rise of cam, in.

y_e = equivalent mechanism follower displacement, in.

y_1 = displacement of follower measured from transition point, in.

y_0 = vertical displacement of prime circle, in.

z = $y_0 + y$ = vertical distance to trace point, in.

A = follower overhang, in.

B = follower bearing length, in.

C = a constant.

D = dynamic magnification factor.

E = distance from cam center to circular arc center of curvature, in.

E_c and E_f = moduli of elasticity of cam and follower respectively, lb/in².

F_a = instantaneous actual force on cam, lb.

F_i = inertia force, lb.

F = force normal to cam profile, lb.

G = maximum follower acceleration, in number of g's.

H = radii of circular arcs, in.

I = moment of inertia, lb-in.-sec².

K_i and K_o = constants for the input and output cams, respectively.

L_0 = total external load on cam (includes weight, spring force, inertia, friction, etc.), lb.

L = external load on follower, lb.

M = equivalent mechanism connecting rod length, in.

N_1, N_2 = forces normal to translating follower stem, lb.

N = cam speed, rpm.

P = pressure, lb/in².

Q = distance between centers, in.

R = function of θ_i.

R_a = radius of prime circle, in.

R_b = radius of base circle, in.

R_g = radius of cutter or grinder, in.

R_p = radius of pitch circle, in.

R_w = mean radius of pin wheel, in.

S = spring force, lb.

S = function of θ_o.

S_c = maximum compressive stress, lb/in.²

S_1 = initial spring force, lb.

S_2 = maximum spring force, lb.

T = torque, lb-in.

T_i = inertia torque, lb/in.

T_0 = bilateral tolerance, in.

U = lever arm or gear ratio.

$V_{P/A}$ = velocity of point P relative to fixed body A, ips.

V_{PN} = velocity of point P normal to contacting surfaces, ips.

$V_{P/Q}$ = velocity of point P relative to point Q, ips.

W = radius of follower stem, in.

X = input cam movement, radians or in.

X = distance to curve in one direction, in.

Y = output follower movement, radians or in.

Y = distance to curve in direction perpendicular to X, in.

α = pressure angle, degrees.

α = angular acceleration, radians/sec².

α_m = maximum pressure angle, degrees.

β = cam angle rotation for total rise h, radians.

β_1 = angle of positive acceleration period, radians.

γ = angle between radius r and tangent to pitch curve, degrees.

δ = compression spring deflection, in.

∂ = maximum difference between straight line and any other curve, in.

ϵ = crank angle rotation (equivalent mechanism) for displacement $y\epsilon$, radians.

η_f	= damping ratio between frame and linkage, lb-sec²/in.
η_l	= damping ratio in follower linkage, lb-sec²/in.
θ	= cam angle rotation for follower displacement y, radians or degrees.
θ_1	= cam angle rotation measured from transition point for follower displacement y, radians.
θ_i	= angle rotated by input driver, degrees.
θ_o	= angle rotated by output driven member, degrees.
θ_p	= cam angle to pitch point, radians.
θ_t	= cam angle to a point on straight-sided flank, radians.
λ	= number of free vibration cycles per positive acceleration period.
μ	= coefficient of friction.
μ_c and μ_f	= Poisson's ratio for cam and follower, respectively.
μ_s	= static coefficient of friction.
π	= 3.14159.
ρ	= radius of curvature, in.
ρ_c	= radius of curvature of cam profile, in.
ρ_f	= radius of curvature of follower face, in.
ρ_k	= radius of curvature of pitch curve, in.
ρ_n	= radius of curvature of cam nose, in.
ρ_i	= radius to point of contact at angle θ_i, in.
ρ_o	= radius to point of contact at angle θ_o, in.
σ	= error or deviation from theoretical cam profile, in.
τ	= angle between connecting rod and follower motion, degrees.
ϕ	= angle of oscillating follower movement for cam angle θ, degrees.
ϕ_o	= total angle of oscillating follower movement, degrees.
φ	= angle of rotation of harmonic-circle vector, radians.
ψ	= angle between initial rise point and radius r, degrees.
ψ_c	= angle between initial rise point and radius r_c, degrees.
ψ_g	= angle between initial rise point and radius r_g, degrees.
ψ_p	= angle between initial rise point and radius r_p, degrees.
ω	= cam angular velocity, radians/sec.
ω_n	= natural frequency of follower linkage, radians/sec.

CHAPTER ONE

Introduction

1.1

Present-day requirements usually dictate that machine components move in a prescribed, exact path. Connected members alone can rarely fulfill this requirement. It is therefore necessary to resort to the use of miscellaneous contour surfaces called cams. Since the cam may have any shape, it is simple and adaptable. Moreover, among the fundamental kinematic linkages the cam has the virtue of being easy to design if properly understood and the action produced by it can be most accurately forecast. To design by use of other mechanisms to produce a given motion, velocity, and acceleration is complicated; to do the same with a cam is by comparison easy, accurate, and efficient. Cams are found in almost all machines, e.g., textile, packaging, machine tools, printing, internal combustion engines, switches, ejection molds, and computers. No one is sure where or how cams got their start. The Sanskrit (Indo-Iranian) term "Jambha" (cog, peg, or tooth) may indicate the geographic area in which they had their beginnings. So may the Teutonic "Kambr" (toothed instrument). Suffice it to say that they have been with us for some time.

A cam is a mechanical member for transmitting a desired motion to a follower by direct contact. The driver is called a cam, and the driven member is called a follower. The cam may remain stationary or translate, oscillate, or rotate, whereas the follower may translate or oscillate. Grodzinski[4] has an excellent compilation of practical cam mechanisms.

Kinematically speaking, in its general form the plain cam mechanism (Fig. 1.1a) consists of two shaped members, A and B, connected by a fixed third body, C. Either body A or body B may be the driver with the other the follower. We may at each instant replace these shaped bodies by an equivalent mechanism having members as shown in

1

Fig. 1.1b. These are pin-jointed at the instantaneous centers of curvature, 1 and 2, of the contacting surfaces. At any other instant, the points 1 and 2 are shifted and the links of the equivalent mechanism have different lengths.

In low-speed, low-weight, cam-follower linkages, the design of the cam may be very simple to construct, requiring little understanding of contours. As either the mass, speed, or elasticity of the follower linkage

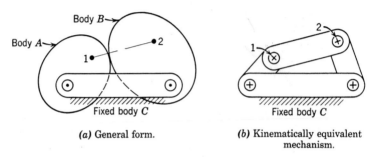

(a) General form.   (b) Kinematically equivalent mechanism.

Fig. 1.1.   Generalized cam mechanism (two shaped bodies in contact).

increases, a detailed study must be made of the contour and its characteristic velocity and acceleration curves. Thus, pertinent information relating to cam size, vibrations, dynamic loads, noise, wear, stresses, etc., is obtained.

The final choice of a cam-follower system is generally an intelligent compromise of the following important factors:

1. The cam almost always runs at a *constant speed* which is established by the desired values of the follower displacement, velocity, and acceleration.

2. The *radial cam* is usually employed.

3. The *cam contour* should be made smooth with no abrupt changes in its curvature.

4. The minimum *curvature* or sharpness of a convex cam contour is dependent upon the value of the maximum negative acceleration of the follower. That is, the larger the maximum negative acceleration, the sharper must be the cam surface.

5. The *cam size* should be as small as possible to minimize the cam-follower sliding velocity, surface wear, and torque. Also, improved balance of the cam will result.

6. The *pressure angle* or steepness of the cam contour should be kept to a minimum.

7. Basic popular curves are available that will fulfill the requirements for most machines (Table 2.2).

8. The follower *acceleration* at high speeds should be as low as possible to keep the inertia forces and stresses small.

9. At high speeds, the *noise, surface wear*, and *vibration* of a cam-
follower system is largely dependent upon the shape of the follower
acceleration curve; smoothness and continuity are essential.

10. Scribing lines on the template or cam blank, as a *cam fabrication
method,* is acceptable only for low-speed action. *Calculated profiles* to
establish exact location for cutter or grinder are necessary for higher
cam speeds

11. *Manufacturing methods* and accuracy of cutting and inspection
are of paramount importance in assuring the anticipated performance.
Required surface and profile accuracy depend upon the cam speed and
the follower displacement. Small surface errors imperceptible to the
eye may produce high stress and vibrations in the follower linkages at
high speeds.

12. The moving parts of the cam-follower mechanism should be made
as *light in weight* and as *rigid* as possible (with no backlash in the
system, especially at high speeds). This helps to keep the inertia forces,
noise, and wear at a minimum.

Followers

1.2   TYPES

Followers in general may be classified in one of three ways: (1) the
construction of surface in contact, e.g., *knife-edge, roller,* or *flat-faced;*
(2) the *type of movement,* whether *translating* or *oscillating;* and (3)
the *location of line of movement* with reference to the cam center, such
as *radial* and *offset.*

Let us now elaborate on these groups. The translating radial follower shown in Fig. 1.2 will be used as an example. The oscillating
follower, as seen in Figs. 1.9 and 1.12, will be discussed later.

The *knife-edge* or point follower (Fig. 1.2a) is, as the name implies,
a sharp knife edge in contact with the cam. Although simple in construction, this type of follower is obviously not practical because it
produces extreme wear of surface and contact point. It is primarily of
academic significance.

A *roller follower* (Fig. 1.2b) is usually cylindrical in shape, moving
on ball or needle bearings held by a pin to the follower assembly.
Crowned and conical rollers have been used to obtain reduced stresses
and wear. The roller action at low speeds is pure rolling, but as the

speed increases more sliding becomes evident. In general, this type of follower is used to reduce considerably the sliding action between the cam and follower. The primary disadvantage of the roller follower is that too steep a cam tends to jam the moving translating follower.

Figures 1.2c and 1.2d show *flat-faced mushroom* followers. Throughout this book, reference to flat-faced followers does not mean that the contact surface is a perfectly flat plane; it may also have a spherical

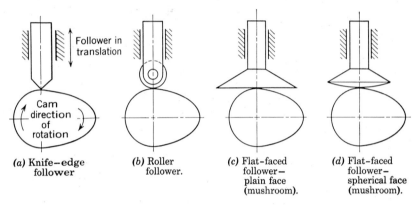

(a) Knife—edge follower   *(b)* Roller follower.   *(c)* Flat-faced follower— plain face (mushroom).   *(d)* Flat-faced follower— spherical face (mushroom).

Fig. 1.2.   Types of follower contact surfaces.   (Translating radial follower is in contact with a radial cam.)

face. The *mushroom follower* of this type has a circular-projected face. The flat-faced follower has an advantage over the roller follower in that excessive steepness of the cam contour produces no impairing effects.

Figure 1.2c shows a flat-faced follower which may give high surface stresses and wear due to deflection and misalignment; to alleviate this condition, we may utilize the spherical face (Fig. 1.2d) having a surface of large spherical radius.   This surface compensates for detrimental deflection and misalignment.

Generally speaking, roller followers are most frequently applied in machinery.   They are available in commercial sizes.   Nevertheless, automobile engines use flat-faced mushroom followers (spherical curvature) in preference to the roller because of limited space, pin weakness, and lubrication difficulty of bearings.   On the other hand, aircraft engines employ roller followers since high cam peripheral velocities preclude the use of any other.   Otherwise, wear would be excessive.

The *radial follower* is one in which the follower translates along an axis passing through the cam center of rotation.   This type, shown in Fig. 1.2, is the most common.

The *offset follower* is one in which the axis of the follower movement is displaced from the cam center of rotation. Offsetting will often improve action by reducing forces and stresses and thus the cam size. The eccentricity should be in the direction to improve force components tending to jam the translating follower in its bearing guide. Figure 1.3*a* shows a follower on a radial cam with an offset shown. In Fig. 1.3*b*, we

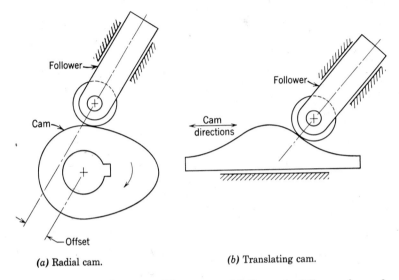

(a) Radial cam.   (b) Translating cam.

Fig. 1.3. Offset followers. (Movement of follower is different from shape of cam.)

see the equivalent relationship for a translating cam. In both cams, contrary to the radial follower, the follower path is not the profile displacement of the cam.

For proper performance, followers must be constrained to the cam at all speeds. This is accomplished either by a *preloaded compression spring* (usually a coiled helical type), a *positive-drive condition,* or by gravity. Sometimes hydraulic or pneumatic means are utilized. Positive-drive action is accomplished by either a roller follower in a cam groove (Fig. 1.5) or a conjugate follower or followers in contact with opposite sides of single cam or dual cams (Figs. 1.8 and 1.9). Experience shows that the grooved cam roller follower does not provide *exact* positive-drive action because of the necessary clearance between the roller sides and its groove. Sliding, wear, noise, vibrations, and shock may be induced at high speeds.

Cam Types

1.3 GENERAL CLASSIFICATION

Cams may be classified in three ways: first in terms of the *follower motion*—such as *dwell-rise-dwell* (*D-R-D*), *dwell-rise-return-dwell* (*D-R-R-D*), or *rise-return-rise* (*R-R-R*); secondly, in terms of their *shape*—such as *wedge, radial, globoidal, cylindrical, conical, spherical, three-dimensional*, and *inverse;* thirdly, in terms of the *manner of constraint* of the follower. This *constraint* is accomplished by either spring loading or positive drive, previously discussed.

Now, in the following articles, we have often shown more than one name to define the types of cams and actions. They are offered principally for reference since all of them have been used. As a best choice, it is suggested that the reader select the first name of a group.

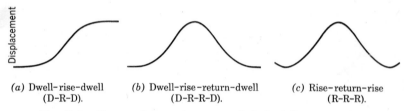

| *(a)* Dwell-rise-dwell | *(b)* Dwell-rise-return-dwell | *(c)* Rise-return-rise |
| (D-R-D). | (D-R-D). | (R-R-R). |

Fig. 1.4.   Types of cams in terms of their follower motion.

1.4 CAMS IN TERMS OF FOLLOWER MOTION

There are three basic types of cams classified in terms of the type of follower motion.

Dwell-Rise-Dwell Cam (*D-R-D*). This type (Fig. 1.4*a*) is one in which part of the cam action is a zero displacement portion called dwell, followed by a rise contour, to another dwell period. This is the most frequent kind in machinery. Into this category also falls the dwell-fall-dwell cam, which is analyzed in a similar manner. Actually, the complete cam is dwell-rise-dwell-fall-dwell action.

Dwell-Rise-Return-Dwell Cam (*D-R-R-D*). This type (Fig. 1.4*b*) has the rise and return preceded and followed by dwells. It has some application in machinery.

Rise-Return-Rise Cam (*R-R-R*). The rise-return-rise contour (Fig. 1.4*c*) has no dwells. Although we shall analyze this type, it will be shown that an eccentric mechanism is suggested (if possible) since it fulfills the follower action more accurately than a cam contour.

In the following articles we shall discuss cams in terms of shapes.

1.5

The *translation, wedge,* or *flat-plate cam* is one which moves back and forth driving a follower. The follower may either translate or oscillate, with its position established by the cam shape and location. This is the simplest cam of all. The follower is held in contact either by a spring or a positive-drive groove and roller (Fig. 1.5). Note that the desired follower movement determines the shape of the groove. Such cams have been built as large as 15 ft long for turning the outside profile on gun barrels for milling or profiling work.

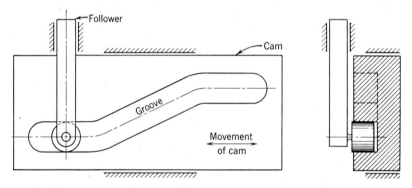

Fig. 1.5. Translating cam. (Follower is a translating roller positive-drive type.)

1.6

The *radial* or *disk cam* is one in which the position of the follower is determined by radial distances from the cam axis. The follower, which translates or oscillates, is held in contact by means of a preloaded compression spring, gravity, or a positive drive. This cam is by far the most popular because of its simplicity and compactness. The following are cams of this type:

The *periphery cam* is one in which the outside surface is in contact with the follower (Figs. 1.7, 1.8, and 1.9).

The *heart cam* is a cam having the shape of a heart.

The *frog cam* is a cam which has the form of several irregular lobes (Fig. 5.4*g*).

The *face-groove* or *plate-groove cam* (Fig. 1.6) is a cam that has a groove cut in the face of the disk. A roller running in this groove constrains the follower to positive-drive action.

The *box cam* is one in which the roller follower is enclosed by two walls. Thus the face cam is one of this category. Sometimes, erroneously, the name box cam is given to the yoke cam later defined.

Fig. 1.6. Radial face-groove cam—employed on hosiery machine. Heat-treated Meehanite cam; turns approximately 50 rpm and drives an extended gear quadrant, which in turn activates other parts of the machine to multiplied higher speeds. Courtesy Wildman Manufacturing Co., Norristown, Pa.

The *wiper cam* is one with a long curved shape driving a curved or flat-faced follower with translation or oscillation. The follower is often called a toe by virtue of its shape. The cam usually oscillates but sometimes may be made to rotate. Figure 1.7 shows the wiper cam with an oscillating follower. Note that the wiper cam slides on the follower surface during its action. If the cam is oscillating, the cam and follower will maintain contact at all times. However, with complete cam rotation, the surfaces will separate and the follower will, by spring action or gravity, have a quick return condition so that the cycle can be repeated. These oscillating cams have been employed as the rocker in steam engine valve gear, and the rotating cam has been used as an ore breaker in stamp mills. This type cam has limited application.

The *rolling cam* is one with pure rolling between the cam and the follower. It is very similar in appearance and application to the wiper cam which is improved by alleviating the detrimental sliding and surface wear. Figure 5.4 shows rolling cams.

Fig. 1.7. Wiper cam. Turning through a small angle, it lifts an exhaust steam valve. Courtesy Skinner Engine Co., Erie, Pa.

The *yoke cam* is a positive-drive cam enclosed by the follower having opposite rollers or surfaces a constant distance apart. The rollers or surfaces may or may not be diametrically opposite each other. The follower may translate or oscillate. Figure 1.8 shows a single-disk surface yoke cam with a translating follower. Control of follower action on single-disk cams is limited to 180 degrees of cam rotation. The other 180 degrees are complementary to maintain the fixed distance contact on the other side of the cam. Another shortcoming of the single-disk yoke cam is that any wear produces clearance, which will result in poor action at moderate to high speeds. Two roller followers may be utilized in lieu of surfaces to reduce somewhat this wear and detrimental backlash (Fig. 4.14).

The *conjugate, complementary,* or *double-disk yoke cam* is one having dual radial disks, each in contact with a roller or surface of the follower. Figure 1.9 shows this type with an oscillating follower. Roller followers are generally used for best performance. Two cams and two followers allow one roller to be preloaded against the other with backlash almost eliminated. This effectively constrains the fol-

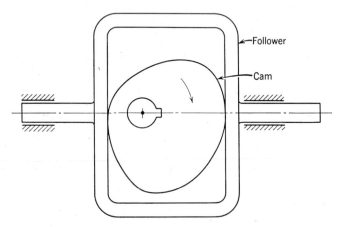

Fig. 1.8. Yoke cam.

lower under high speed and high dynamic or shock loads. Low noise, vibration, and wear, and excellent control of the follower result. With this cam, complete follower control of 360 degrees of cam rotation may be achieved in contrast with the limited 180-degree control of the single-disk yoke cam.

The *circular arc cam,* as the name implies, is one composed of circular arcs. This cam is usually a radial type. Figure 5.12 shows a circular arc cam with a translating roller follower. The advantages of the circular arc cam are simplicity of construction, ease of cutting in small quantities, and ease of checking the contour accuracy.

The *harmonic cam* is a type of circular arc cam giving partial or complete simple harmonic motion to the follower. The simple harmonic motion will be exact for a translating follower and approximate for an oscillating one. This cam is usually a form of single-disk yoke cam having two or more enclosing sides. The harmonic cam may move the follower in geometrical paths, such as triangular, square, pentagonal, etc. In this manner, holes of these shapes have been successfully drilled (Fig. 11.9).

The *spiral cam* is a form of face cam having a special spiral groove which drives a translating follower (by having a pin in the groove) or a rotating follower. The action may occur in more than 360 degrees of cam rotation. In Fig. 1.10, the pin gear follower is driven by teeth in the cam groove. The follower angular velocity is determined by the

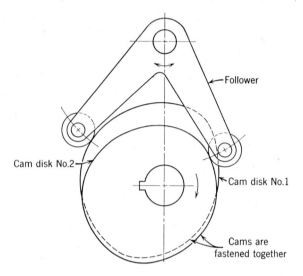

Fig. 1.9. Conjugate cam—oscillating follower. (Dual rollers and cams.)

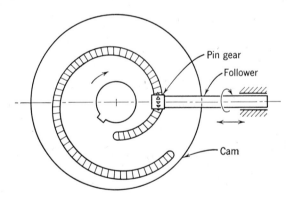

Fig. 1.10. Spiral cam. (Pin gear meshes with teeth in cam groove.)

radial distance from the groove to the cam axis. The spiral cam is often applied to computer mechanisms. In this application, which uses the Archimedes spiral, the angle of follower rotation can be shown to be the square of the cam angle of rotation.

The *internal cam* is a radial cam having its actual cam profile or contact surface radially outside the roller or surface follower. This cam is analogous to annular or internal gears. The follower is held in contact by spring or gravity. The primary advantage of the internal cam is that a slightly larger roller diameter is permitted than with the popular peripheral type.

1.7

The *globoidal* or *barrel cam*, rotating about its axis, has a circumferential contour cut into a surface of rotation. The surface shape is determined by the arcs of an oscillating lever follower. The globoidal cam is either convex (Fig. 1.11*a*), or concave (Fig. 1.11*b*). Usually

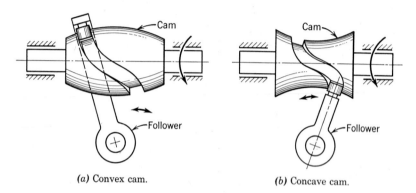

(*a*) Convex cam.   (*b*) Concave cam.

Fig. 1.11. Globoidal cam—oscillating follower.

this cam can be replaced by the more practical cylindrical cam (later defined) if the angle of oscillation is small. As shown in Fig. 11.24, this globoidal cam may be utilized for indexing an intermittent rotating follower. In all cams of this type, necessary backlash in the roller follower groove will affect the accuracy and performance, especially under high speeds and high loads.

1.8

The *cylindrical, barrel,* or *drum cam* has a circumferential contour cut in the surface of a cylinder which rotates about its axis. The follower is oscillated or translated in the direction of this axis. The cylindrical cam in essence is similar to the globoidal cam. The difference is in their basic shapes. With cylindrical cams, cylindrical roller followers are usually used, with conical roller followers as an alternate choice for small cams. The cylindrical cam has been employed internally as well as externally as follows:

The *plain cylindrical cam,* commonly called cylindrical cam (Fig. 1.12) is one with a groove cut in its surface to constrain a roller follower in a desired motion. The plain cylindrical cam is a positive-drive type which is second in popularity to the radial-disk cam.

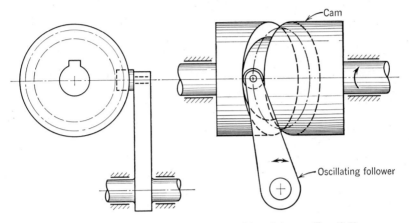

Fig. 1.12. Plain cylindrical cam—positive-drive roller follower.

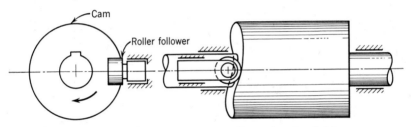

Fig. 1.13. End cam. (Usually spring loaded.)

The *end* or *bell cam* (Fig. 1.13) has as its working surface the end of a cylinder. This cam requires a spring or gravity to maintain contact with its follower.

1.9

The *conical cam* (Fig. 1.14) has the shape of a cone driving the follower in a plane perpendicular to the base of the cone. The follower translates or oscillates whereas the cam usually rotates. This cam is used very infrequently because of cost and difficulty in cutting its contour. However, the conical cam may be necessary to change the direction of motion in a limited space. Grooves have been cut in the conical cam for positive-drive action of the follower. In this application the conical cam has been employed in operating the valve in the Herrmann internal combustion engine.

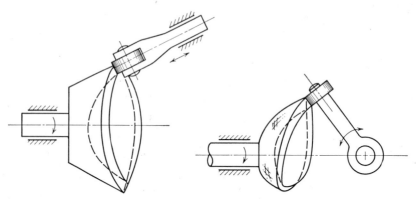

Fig. 1.14.   Conical cam—trans-
lating roller follower.

Fig. 1.15.   Spherical cam—oscillating
roller follower.

Fig. 1.16.   Three-dimensional cam—Meehanite cams used in the manufac-
ture of the solid aluminum blade shown.   Employed as the master in cutting
the proper contours in the face and side of the blade.   Courtesy Hamilton
Standard Division, United Aircraft Corp., East Hartford, Conn.

1.10

The *spherical cam* (Fig. 1.15) has a rotating portion of a sphere which oscillates a follower having its axis perpendicular to the cam axis. This cam is the limiting case of a convex globoidal cam. The realms and restrictions of the spherical cam are similar to the conical cam. Therefore, it should not be employed unless definitely necessary to a particular design.

1.11

The *three-dimensional cam,* or *camoid* (Fig. 1.16) having a curved face, is both rotated about the longitudinal axis and moved relative to the follower along this axis. Thus the position of the translating follower is dependent upon these two variables: the cam angular and translational positions. The three-dimensional cam has limited application because of its high fabricating cost, high frictional forces that may be induced, and large space requirements. Although this cam usually is applied as a last resort, there are designs that definitely justify its application—computer mechanisms are another example.

1.12

The *inverse cam* is one in which the element corresponding to the follower of a cam mechanism is utilized as the driver. Inverse cam mechanisms are rarely used. Figure 1.17 shows this cam mechanism with an oscillating roller driving a translating follower having a curved face. If the profile were a straight path perpendicular to the direction of follower motion, the familiar Scotch yoke mechanism would result.

1.13 ADDITIONAL DEFINITIONS

The following are some miscellaneous definitions that are applied to cams:

Stationary cams. Some cams are fixed and are machined with a curve or contour calculated to produce auxiliary motions on moving parts contacting them during traversal. These are generally wedge cams and are used infrequently. They have been employed in high-speed wrapping machines.

The *double-end cam,* or *roller gear drive,* is one in which the cam has a projected, twisted cam thread in contact with a two-roller follower, one on each side of the thread. The follower may translate, oscillate, or have intermittent rotation. The body of the cam may be either cylindrical or globoidal, depending on the follower location and movement. Figure 1.18 shows a single-thread type with a globoidal body and an oscillating follower. These cams have high accuracy, long life,

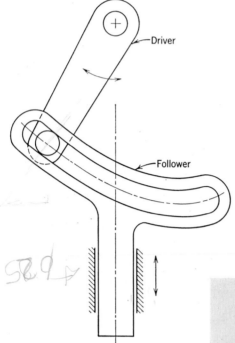

Fig. 1.17. Inverse cam mechanism.

low vibration, low noise, and shockless performance under high speeds. For best action, it is suggested that the bearings should be pre-loaded so that the system has limited backlash. This cam and the follower can be obtained commercially. For indexing, they are often substituted for the familiar Geneva and star wheel mechanisms which have inherently poor high-speed dynamic properties.

Step cams. All cams are step cams. A single-step cam is one in which the total follower rise is made in a single continuous movement or step. All previous cams are single-step cams. Multiple-step cams such as the double and triple are rarely utilized in practice.

The *adjustable, clamp, strap, dog,* or *carrier cam* is one having the cam shape of formed, removable pieces bolted to a supporting member. If these pieces are attached to a rotating disk, we have radial cam action. On the other hand, putting these pieces on a cylinder will give cylin-

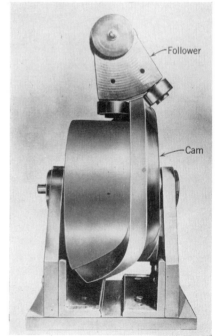

Fig. 1.18. Double-end cam—high-speed positive-drive oscillating follower. Courtesy Ferguson Machine and Tool Co., St. Louis, Mo.

drical cam action. The adjustable cams offer quick changes to other contours. One of the applications of the cylindrical dog cam is in stock feeding for automatic machine lathes (Fig. 1.19). Dogs are also fastened to chains to control the pattern in knitting machines.

Fig. 1.19. Adjustable cylindrical cam. Courtesy Jones and Lamson Machine Co., Springfield, Vt.

The *master, leader, model,* or *former cam* is used to guide the cutter or grinder in fabricating high-production cams. It is generally oversized for increased accuracy and hardened for long wear life. The master cam is made as precise as cost and time will permit.

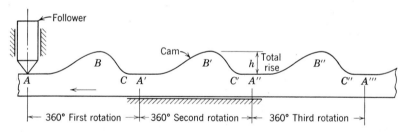

Fig. 1.20. Wedge cam—equivalent to all cam actions.

Empirical Cam Design

1.14 DISCUSSION

Cam action may be considered equivalent to wedge action (Fig. 1.20). Similar to the cam, the wedge is shown moving at a constant speed from right to left to raise the follower a maximum distance h. The first cycle of follower movement is the curve $ABCA'$ continued to $B'C'A''B''C''A'''B'''$, etc. However, this action repeated many times would require a cam of impractical length. Therefore, for a reasonable cam size, a radial or cylindrical cam is common as an equivalent to a wedge cam of infinite length. A radial cam may be considered a wedge rolled in the plane of the paper having a length of one cycle wedge cam $ABCA'$ equal to the circumference. A cylindrical cam is one in which the wedge $ABCA'$ is twisted in a plane perpendicular to the paper with the length of wedge cam $ABCA'$ equal to the circumference of the cylinder. Further reference to Fig. 1.20 shows the following points of interest:

1. The contour AB, which is the rise portion of the cam, should not be too steep since this will produce jamming of the translating follower on the sides. This curve slope can be reduced by using a longer cam length for the same rise of the follower h. A larger cam results.

2. The contour $ABCA'$ should be smooth, free of sharp changes or "bumps."

3. The speed of the wedge cam is pertinent to later investigation of cam-follower action—wear, shock, spring size, dynamic loads, vibration, and lubrication all being affected.

At this point in our study, we can construct a simple cam requiring a given total rise in a given time. If the cam speed is very low, no limitations other than that the cam should have a smooth contour and be large enough need to be considered. This is often called the *empirical method* for cam design. Throughout this book we shall present means for controlling and understanding the cam action essential to the usual designs.

The basic method for the construction of a cam contour is:

The profile is developed by fixing the cam and moving the follower around the cam at its respective relative positions.

If we took an infinite number of points, the envelope of the cam contour would be formed. In cam layout, it is necessary to draw enough points for a smooth, reliable cam contour. Accuracy is essential. Also, note that the layout of a roller follower requires plotting the center of the roller and then drawing the curve tangent to the rollers. In the following example, let us use the empirical method for the layout of a low-speed cam. For clarity, the location of only a few points will be shown.

Example

A radial cam rotating clockwise drives a ½-inch-diameter roller follower as follows: (*a*) rise of ⅝ in. in 90 degrees of cam rotation;

(b) fall of ⅝ in. in 90 degrees of cam rotation; (c) no follower movement (dwell) for the last 180 degrees of cam cycle.

The procedure (Fig. 1.21) which consists of fixing the cam and moving the follower around it is:

(a) Draw two perpendicular axes locating the cam center at A.

(b) Assume the location of the lowest point of the roller-follower center at point 0, on one of the radial lines.

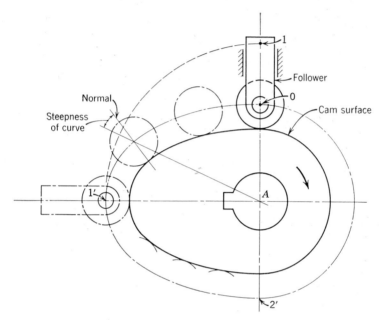

Fig. 1.21.   Cam layout—empirical method.   Rise is ⅝ in. in 90 degrees of cam rotation.   Full scale.

(c) The maximum rise, point 1, should be shown ⅝ in. above point 0.

(d) With the cam center A and radius A1, swing an arc until it intersects line A1′, which is 90 degrees opposite the direction of cam rotation.

(e) Draw a smooth curve through points 0 and 1′, giving the rise portion of the cam.

(f) Draw roller diameters and then a tangent curve to the rollers. This is the cam surface.

(g) Check steepness of curve—if at any point it is more than 30 degrees, we must redesign with larger radii A0 and A1, which means a larger cam.

(h) Repeat the process for the 90-degree fall portion 1′2′. The last 180 degrees of cam action 2′0 is at a constant radial distance.

Thus we see that a low-speed cam requires a "bumpless" curve that is not too steep. No advanced study is necessary with this introductory empirical trial-and-error method.

References

1. F. DeR. Furman, *Cams, Elementary & Advanced*, John Wiley & Sons, New York, 1921.
2. F. B. Jacobs, *Cam Design and Manufacturing*, D. Van Nostrand, New York, May, 1921.
3. L. Kasper, *Cams, Design and Layout*, First Ed., Chemical Publishing Co. Inc., New York, 1954.
4. P. Grodzinski, *A Practical Theory of Mechanisms*, First Ed., Emmott and Co. Ltd., Manchester, England, 1947.
5. I. I. Artobolevskii, *Theory of Mechanisms and Machines* (in Russian) Chief State Publishing House for Technical Theoretical Literature, Moscow, USSR, 1953.
6. F. P. Jones, *Ingenious Mechanisms for Designers and Inventors*, First Ed., Industrial Press, New York, Vol. 1, p. 1, 1948; Vol. 2, p. 1, 1948; Vol. 3, p. 1, 1951.
7. F. Reuleaux, *The Kinematics of Machinery*, translated by A. B. W. Kennedy, Macmillan, London, 1876.

C H A P T E R   T W O

Basic Curves

2.1 INTRODUCTION

We have seen in the previous chapter that it was possible to construct a simple cam by its appearance to the eye. However, this method provides no information as to the follower velocity and acceleration. These factors are of pertinent concern in most machines since proper design requires accurate analysis and control of the velocity and acceleration curves. In general, the higher the speed, the more critical becomes the investigation. This is especially true of the acceleration curve which is the determining factor of the dynamic loads and vibrations of a cam-follower system.

For years mathematically related *basic curves* have been found convenient. These curves have proved popular from a standpoint of ease of layout, reproduction analysis, and control. In this chapter, we shall present the methods of construction and the characteristic equations of the basic curves. For a mathematical verification of these curves under high-speed action, the reader is referred to Chapter **8**.

2.2 DEFINITIONS

First, let us define the terms necessary for the investigation of the cam shapes and actions.

The *displacement diagram* (Fig. 2.1) is a rectangular coordinate layout of the follower motion in one cycle of cam operation. The rise of the follower is shown as the ordinate with the length of the abscissa arbitrarily chosen. The abscissa is divided into equal cam angles or equal time divisions since the cam usually rotates at a constant speed.

The displacement diagram is generally drawn or sketched as the first step in the development of the cam profile. It presents a quick picture of follower motion. Later it will be shown that for radial cams the displacement diagram, if drawn to scale, does not show the true slopes of the cam contour. In other words, in transferring the cam contour from a rectangular to a polar layout the displacement diagram is distorted on radial cams whereas no change occurs with cylindrical cams.

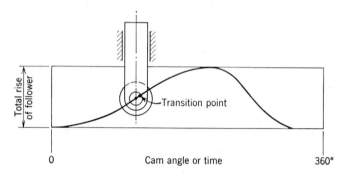

Fig. 2.1.   Displacement diagram (shows cam action).

The *transition point* (Fig. 2.1) is the point on the cam at which the follower has its maximum velocity. In the displacement diagram, the transition point (point of inflection) is located at the maximum cam slope.

The *time chart* (Fig. 2.2) is the superimposing of more than one displacement diagram on the same abscissa or time basis. This pro-

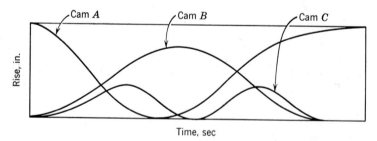

Fig. 2.2.   Time chart (showing compounding of three cams).

vides a comparison of the operation of interrelated cams. The plotting of this chart is essential in automatic machinery to prevent interference of the cam followers and to maintain proper timing of the members. Therefore, by proper use of this chart, the designer can keep the idle time to a minimum and increase the production of the machine. Figure 2.2 shows the time charts used for the compounding of three cams. The usual development of high-speed automatic machinery requires the manipulation of the timing diagram until the ultimate design has been reached.[1]

The *cam profile* is the actual working surface contour of the cam. It is the surface in contact with the knife-edge, roller surface, or flat-faced follower. Figure 2.3 shows a popular cam profile consisting of a single-lobe, external radial cam. The cam profile may be any shape whatsoever, external or internal, single- or multi-lobe, etc.

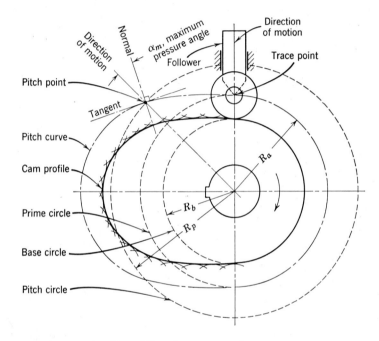

Fig. 2.3.   Cam nomenclature.

The *base circle* (Fig. 2.3) is the smallest circle drawn to the cam profile from the radial cam center. Obviously, the cam size is dependent upon the established size of the base circle. In the empirical design method of Art. 1.14, the base circle was first assumed and then the cam was constructed. We shall denote the radius of the base circle in inches as R_b.

The *trace point* (Fig. 2.3) is the point on the follower located at the knife edge, roller center, or spherical-faced center.

The *pitch curve* is the path of the trace point. Figure 2.3 shows the pitch curve of a radial cam. In cam layout, this curve is often determined first with the cam profile established by tangents to the roller or flat-faced follower surfaces. For the elementary knife-edge follower, the pitch curve and the cam profile are colinear.

The *prime circle* (Fig. 2.3) is the smallest circle drawn to the pitch curve from the cam center. It is similar to the base circle. We shall denote the radius of the prime circle in inches as R_a.

The *pressure angle* (Fig. 2.3) is the angle (at any point) between the normal to the pitch curve and the instantaneous direction of the fol-

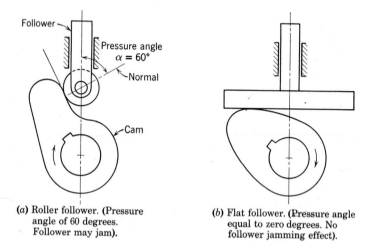

(a) Roller follower. (Pressure angle of 60 degrees. Follower may jam).

(b) Flat follower. (Pressure angle equal to zero degrees. No follower jamming effect).

Fig. 2.4. Significance of pressure angle.

lower motion. This angle is important in cam design since it represents the steepness of the cam profile which, if too large, can affect the smoothness of the action.

Suppose that, as in Fig. 2.4a, the cam had a pressure angle of approximately 60 degrees. It can be seen that there is a strong possibility that clockwise rotation of the cam would not cause the translating follower to rise but to jam against the guides and cause bending in the follower stem. Designers often empirically limit the pressure angle to 30 degrees or less for smooth cam-follower action. However, if the follower bearings are good, the cam follower is rigid, and the cam-follower overhang is small, the maximum pressure angle may be increased to more than 30 degrees. In the flat-faced followers this locking action is not as serious because this follower has a small, constant-pressure angle. In Fig. 2.4b, we see a follower face normal to the translating follower motion in which the pressure angle is constant at zero degrees. Thus, no jamming occurs. Let us establish the following: α = pressure angle, in degrees; α_m = maximum pressure angle, in degrees.

The *pitch point* (Fig. 2.3) is the point on the cam pitch curve having the maximum pressure angle, α_m. The pitch point is significant in cam design since it is the start of the construction of the smallest cam in contact with a roller follower. For cylindrical cams, it is also the point on the displacement diagram having the maximum slope. In other words, the pitch point and transition point are at the same place. However, this is not perfectly true for radial cams, in which the slope changes in going from the displacement diagram to the cam layout. This factor is important in the determination of the exact pressure angle.

The *pitch circle* is one with its center at the cam axis passing through the pitch point. The pitch circle serves no purpose other than that of obtaining a cam of minimum size for a given pressure angle with a translating roller or knife-edge follower. Figure 2.3 shows the pitch circle of a radial cam. With cylindrical cams, the pitch circle is generally at the mean radius of the roller length (Fig. 4.16). The radius of the pitch circle in inches is R_p.

2.3 FOLLOWER CHARACTERISTICS

As previously indicated, a cam could be considered similar to a wedge having a rise and fall which establish the follower motion. In all cams, the displacement of the follower is given by the mathematical relationship

$$y = f(\theta) \quad \text{in.} \tag{2.1}$$

where θ = cam angle rotation in radians. However, since the cam rotates at a constant angular velocity, the displacement

$$y = g(t) \quad \text{in.} \tag{2.2}$$

and

$$\theta = \omega t \tag{2.3}$$

where t = time for cam to rotate through angle θ, sec.

ω = cam angular velocity, rad/sec.

Equations for the follower action are often preferred in the form of eq. 2.1 in lieu of eq. 2.2, since it is simpler to analyze and use.

Throughout the book, unless stated otherwise, we shall establish the *velocity* as the instantaneous time rate of change of displacement

$v = dy/dt$ = slope of the displacement curve at angle θ or time t, ips

The *acceleration* being the instantaneous time rate of change of velocity,

$a = d^2y/dt^2 = dv/dt$

= slope of the velocity curve at angle θ or time t, in./sec^2

The shape and values of the acceleration curves are of critical concern for the designer of moderate- to high-speed machinery. From it, analysis can be made for the shock, noise, wear, vibrations, and general performance of a cam-follower system. It will be shown later that for best action the acceleration curve shall be smooth and have the smallest maximum values possible.

We shall establish another term *pulse* (commonly called jerk) to define the instantaneous time rate of change of acceleration

$p = d^3y/dt^3 = da/dt$
= slope of the acceleration curve at angle θ or time t, in./sec^3

For high-speed actions, it will be shown later that the maximum values of the pulse should not be too large. Vibrations will then be kept to a minimum.

Usually the cam profile is given as a function of angle θ in the form of eq. 2.1. Therefore, to find the velocity and acceleration with respect to time,

$$v = dy/dt = d\theta/dt \cdot dy/d\theta = \omega(dy/d\theta) \quad \text{ips} \qquad (2.4)$$

and

$$a = d^2y/dt^2 = d^2\theta/dt^2 \cdot d^2y/d\theta^2$$
$$= \omega^2(d^2y/d\theta^2) \quad \text{in./sec}^2 \qquad (2.5)$$

These equations facilitate converting from one family of dimensional units to another.

Having the cam profile in mathematical form of eq. 2.1 or eq. 2.2, we can, by differentiation, easily find the other characteristics. It is the mathematical approach used in this chapter. Occasionally the differentiation is cumbersome or the cam displacement curve has no relationship and is in numerical or graphical form. We have seen that each successive derivative can be determined by the slopes of the previous curve. Therefore, to find the velocity, acceleration, or pulse, we may resort to the *graphical-slope differentiating method*. It should be noted that this method requires the utmost care and large scale to give values that are reasonably reliable. Special caution is further suggested at the transition points in the curves. With the graphical-slope method, the velocity curve is reasonably accurate and the acceleration curve is at best only approximate. Therefore, this approach is suggested only as a last resort. Equations and their derivatives are preferred and should be employed where possible.

With the graphical-slope method, the determination of the signs (positive or negative) is essential for a fundamental understanding of the follower motion and dynamics. Let us establish as positive $(+)$

the displacement of the follower above the lowest point in the displacement diagram (Fig. 2.5a). Therefore, the velocity is positive if the
follower moves in the direction of the positive displacement and
negative if it moves in the opposite direction. That is, velocity is
positive on the rise and negative on the fall. The acceleration follows

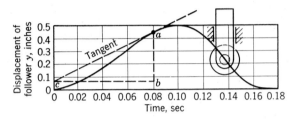

(a) Displacement diagram. (Slope
of curve gives velocity values).

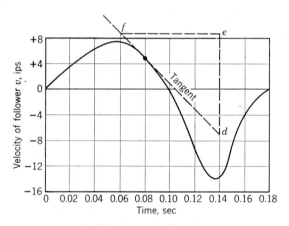

(b) Velocity curve. (Slope of curve
gives acceleration values).

Fig. 2.5. Graphical-slope method for follower characteristics.

a similar convention. Therefore the acceleration is positive when
its direction is that of the positive displacement. Throughout this
book the word deceleration is not mentioned since it would not indicate the direction of acceleration and would therefore confuse the
direction of inertia load. The slope of the curves at any point gives
us another method for determining signs. In the first quadrant, the
slope is positive, and, in the second quadrant, negative.

Example

Given the displacement diagram of a pitch curve in Fig. 2.5a, in which the follower has a total rise of ½ in. in 0.1 sec of cam rotation and falls in an additional 0.08 sec. Plot the velocity curve, and determine the velocity and acceleration after the initial 0.08 sec of cam rotation by applying the graphical-slope differentiating method.

Solution

At the 0.08 sec point, we draw a tangent to the displacement curve. Caution is suggested. This is the instantaneous slope of the curve which equals the velocity.

$$v = dy/dt$$

$$= \text{distance } ab/\text{distance } bc$$

$$= \frac{0.371}{0.08}$$

$$= 4.64 \text{ ips}$$

Note that distances ab and bc were arbitrarily chosen for reasonable measurement accuracy. In the same manner, other velocity points may be found and the curve plotted as in Fig. 2.5b. The acceleration at 0.08 sec is

$$a = dv/dt$$

$$= \text{distance } de/\text{distance } ef$$

$$= -\frac{15.8}{0.08}$$

$$= -198 \text{ in./sec}^2$$

Again, the distances are chosen arbitrarily. Note, the minus acceleration exists because the follower is increasing its acceleration in a minus, downward direction. Also, mathematically speaking, the slope is to the left (second quadrant), which means a negative value. The pulse values may be found in the same manner by plotting the acceleration curve and finding the slopes.

2.4 BASIC CURVE CLASSIFICATION

The basic curves of the rise-fall displacement diagram are primarily of two families: the *simple polynomial* and the *trigonometric*. The trigonometric curves are superior to the polynomial curves and give smoother action, easier layout, lower manufacturing cost, and less vibration, wear, stresses, noise, and torque, in addition to smaller cams.

Simple polynomial curves

The displacement equations of simple polynomial curves are of the form

$$y = C\theta^n \tag{2.6}$$

where n = any number.

C = a constant.

In this polynomial family, we have the following popular curves with integer powers: *straight line*, $n = 1$; *parabolic or constant acceleration*, $n = 2$; *cubic or constant pulse*, $n = 3$.

Trigonometric curves

The curves of trigonometric form are: *simple harmonic motion* (*SHM*) or *crank curve*, which has a cosine acceleration curve; *cycloidal*, which has a sine acceleration curve; *double harmonic*; and *elliptical*.

Other curves

In addition to these two families, we have the miscellaneous little-used curves: *modified straight-line circular arc* and the *circular arc* curves. These are employed primarily as an improvement over the characteristics of the straight-line curve.

Example

A cam rotating at 120 rpm has the positive acceleration part of its rise of $\frac{3}{8}$ in. in 40 degrees of cam rotation. A simple polynomial curve having $n = 2.4$ is used. Find the velocity and acceleration values at the end of 30 degrees of cam rotation.

Solution

The angular velocity of cam = $\omega = 120 \times \dfrac{2\pi}{60} = 12.56$ rad/sec. The total cam angle = $40\pi/180 = 0.698$ radian. Substituting into eq. 2.6 gives the displacement

$$y = C\theta^n = \tfrac{3}{8} = C(0.698)^{2.4}$$

Solving yields $C = 0.889$. Therefore, the basic equation is

$$y = 0.889\theta^{2.4}$$

The velocity by differentiating eq. 2.6 is

$$v = \frac{dy}{dt} = C\omega n\theta^{n-1} \tag{2.7}$$

Thus, the velocity after 30 degrees of rotation is

$$v = 0.889(12.56)(2.4)\left(30 \times \frac{\pi}{180}\right)^{1.4} = 10.8 \text{ ips}$$

The acceleration, differentiating again, is

$$a = \frac{dv}{dt} = C\omega^2 n(n-1)\theta^{n-2} \tag{2.8}$$

And the acceleration after 30 degrees of rotation is

$$a = 0.889(12.56)^2(2.4)(1.4)\left(30 \times \frac{\pi}{180}\right)^{0.4} = 364 \text{ in./sec}^2$$

Let us now establish the method of layout and the mathematical relationships of all curves, first with the dwell-rise-dwell cam. In all the illustrative examples shown in each article, we shall use the same data so that meaningful comparisons can be made. Unless stated otherwise, the curves are *symmetrical about their transition points during rise or return action.* For unsymmetrical curves, the reader is referred to Chapter 6. Last, it is emphasized that, for consistency and minimum error, the cam angle θ is used mathematically in its dimensionless form of radians. If θ has the unit of degrees, it will be specifically stated.

2.5 STRAIGHT-LINE, UNIFORM DISPLACEMENT, OR CONSTANT-VELOCITY CURVE (DWELL-RISE-DWELL CAM)

This curve of the polynomial family ($n = 1$) is the simplest of all. It has a straight-line displacement curve at a constant slope (Fig. 2.6) giving the smallest length for a given rise of all the basic curves. We see that the displacement is uniform, the velocity is constant, and the acceleration is zero during the rise. But, at the ends where the dwell meets this curve, we have an impractical condition. That is, as we go from the dwell (zero velocity) to a finite velocity we have an instantaneous change in velocity, giving a theoretically infinite acceleration. This acceleration transmits a high shock throughout the follower linkage—the magnitude depending on its flexibility. In other words, we have a "bump" in the contour which neither a roller nor other follower could follow. With a dwell-rise-dwell cam, this curve is therefore not practical.

When the straight-line curve is laid out for a radial cam it becomes the *Archimedes spiral*, see Art. 5.5. It is the curve for the "heart" cam which converts rotary motion into uniform reciprocal motion, utilized in automatic machine tools where the cutters are to be moved in a constant rate at extremely low speeds. In this feasible application, no

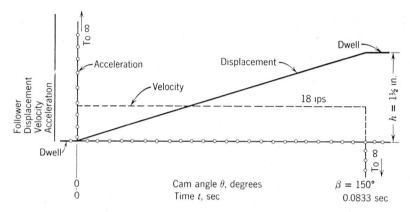

Fig. 2.6. Characteristics of straight-line curve (D-R-D). (Follower has 1½-in. rise in 150 degrees of cam rotation at 300 rpm.)

dwell periods are contained. This curve has frequently been employed in combination with others.

Characteristics

Let h = maximum displacement of follower, in.

  β = cam angle of rotation for rise h, radians.

Using eq. 2.6, we find the displacement of follower

$$y = h\theta/\beta \quad \text{in.} \tag{2.9}$$

Differentiating in the range of the curve for the velocity and acceleration, we find

$$v = dy/dt = h\omega/\beta \quad \text{ips} \tag{2.10}$$

$$= \text{a constant}$$

$$a = dv/dt = 0$$

In Fig. 2.6, we have plotted the characteristic curves of a cam turning 300 rpm, giving the follower 1½ in. straight-line rise in 150 degrees of cam rotation. The calculations will not be set forth since they are similar to the example of Art. 2.4.

2.6   STRAIGHT-LINE CIRCULAR ARC CURVE (DWELL-RISE-DWELL CAM)

To improve the poor condition of sharp bumps on straight-line cams, we can smooth out the junction between the dwells and the rise. This is often achieved by employing circular arcs tangent to both straight-line rise and dwell curves. Usually, the circular arc is formed by a radius equal to the total rise h (Fig. 2.7). The shorter the radius, the nearer is the approach to the undesirable condition of the straight-line curve. A longer radius produces a more gradual action at the beginning and at the end of the curve. Although such a curve is an improvement over the straight-line curves, it can be applied for low speeds only, since large accelerations exist at the beginning and at the end of the stroke.

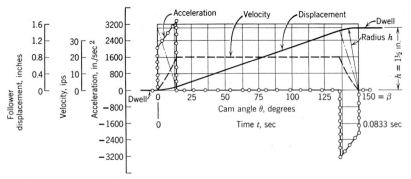

Fig. 2.7.   Characteristics of straight-line circular arc curve (D-R-D). (Follower has 1½-in. rise in 150 degrees of cam rotation at 300 rpm.)

Again, the data for the example are: A cam rotates at 300 rpm with the follower being driven 1½ in. with a straight-line circular arc curve in 150 degrees of cam rotation. The displacement curve is plotted in Fig. 2.7 with the other characteristics found by the graphical-slope method of Art. 2.3. In view of the poor characteristics of this curve, mathematical presentation is omitted. It is rarely utilized and then only at low speeds.

2.7   CIRCULAR ARC CURVE

This curve (Fig. 2.8) composed of two circular arcs tangent to each other is only acceptable insofar as it has some improvement over the infinite acceleration, straight-line curve. Although its acceleration is finite at all times, the curve gives large follower accelerations and excessive velocities. Therefore, it is used for low speeds only, if at all.

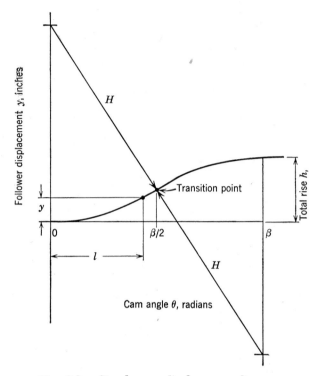

Fig. 2.8.   Circular arc displacement diagram.

Characteristics

Let H = radii of circular arcs, in.

$\quad R_p$ = radius of pitch circle, in.

$\quad l$ = developed length of cam for angle θ, in.

By inspection, we see that the displacement of the follower from 0 to $\beta/2$ is

$$y = H - [H^2 - l^2]^{1/2} \quad \text{in.} \tag{2.11}$$

But we know for radial cams

$$l = R_p\theta \tag{2.12}$$

Substituting gives displacement

$$y = H - [H^2 - (R_p\theta)^2]^{1/2} \quad \text{in.} \tag{2.13}$$

Differentiating gives velocity and acceleration

$$v = dy/dt = \frac{\omega R_p^2\theta}{[H^2 - (R_p\theta)^2]^{1/2}} \quad \text{ips} \tag{2.14}$$

$$a = dv/dt = \frac{(\omega R_p H)^2}{[H^2 - (R_p\theta)^2]^{3/2}} \quad \text{in./sec}^2 \tag{2.15}$$

This curve is symmetrical about the transition point $\beta/2$. No calculated example will be shown since this curve is not employed in modern machinery. However, circular arc characteristic curves are plotted in

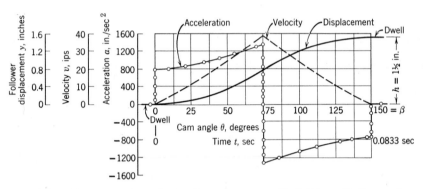

Fig. 2.9. Characteristics of circular arc curve (D-R-D). (Follower has 1½-in. rise in 150 degrees of cam rotation at 300 rpm. $H = 4.62$ in., $R_p = 1.90$ in.)

Fig. 2.9 (comparable to the other articles for reference) with the following data: A circular arc cam of radii $H = 4.62$ in. and pitch-circle radius, $R_p = 1.90$ in., rotates at a constant speed of 300 rpm. The total rise is 1½ in. in 150 degrees of cam rotation.

2.8 ELLIPTICAL CURVE (DWELL-RISE-DWELL CAM)

This member of the trigonometric family is developed from projections of a semiellipse (Fig. 2.10). The contour of the elliptical curve and its characteristics depend upon the assumed proportions of the major and minor axes. As the horizontal axis increases the cam becomes larger with the velocities of start and stop slower. In other words, the curve is flatter at the top and the bottom as the ratio of the horizontal axis to the vertical axis is made larger. If the horizontal axis of the ellipse is zero in length, the contour in the displacement diagram is a straight-line curve. A ratio of $2:4$ gives a small cam for a given pressure angle. Increasing the ratio further to $11:8$ makes the curve approach a parabolic curve, discussed later. At this ratio, a fair cam-follower performance can be expected at moderate cam speeds. Further increase in the ratio is not practical, since velocity, acceleration, and cam size become prohibitive.

One of the primary shortcomings of the elliptical curve is that mathematical treatment is complex. Therefore, other curves are generally preferred because they permit better contour control and ease of duplication and analysis. For equations of the elliptical contour, see Appendix A.1.

Construction

The layout consists of the projections of equal arcs on an ellipse of assumed major to minor axis ratio (Fig. 2.10).

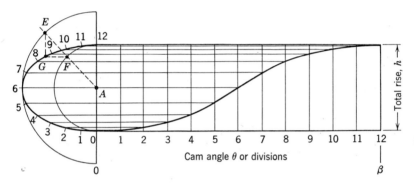

Fig. 2.10. Elliptical curve construction.

(*a*) Plot ordinate and abscissa axes with the total rise of follower *h* equal to either the major or minor axis of the ellipse.

(*b*) Describe 2 circles whose diameters are equal to major and minor axes.

(*c*) Draw any radius *AE* cutting circles at *E* and *F*.

(*d*) Draw lines through *E* parallel to one axis and through *F* parallel to the other. The intersection *G* of two lines is a point on the ellipse. Continue in this manner, and draw the ellipse.

(*e*) Divide both the abscissa of the displacement diagram and the arc of the ellipse into the same number of equal parts, usually 4, 6, 8, 10, 12, or 16. Note, no relation exists between the lengths of divisions on the displacement diagram and the divisions on the elliptical arc except that they must be the same in number.

(*f*) Project these intercepts to their respective cam angle division, and connect points to yield the curve.

In view of the arbitrary major-minor axis ratio chosen and also because of the fact that this curve is rarely used, mathematical relationships and their curves will not be presented. If needed, the velocity and acceleration can easily be found by the graphical-slope differentiation method of Art. 2.3.

2.9   SIMPLE HARMONIC MOTION OR CRANK CURVE (SHM) (DWELL-RISE-DWELL CAM)

This curve of the trigonometric family is one of the most popular primarily because of its simplicity in layout and understanding. It provides acceptable performance at moderate speeds. The basis for the harmonic curve is the projection (on a diameter) of the constant angular velocity movement of a point on the circumference of a circle. This circle is called the harmonic circle. The resulting motion of the

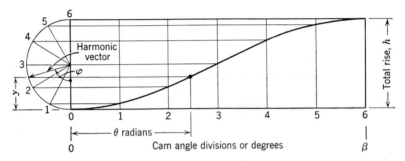

Fig. 2.11.   Simple harmonic motion (SHM) construction.

follower on such a cam is simple harmonic movement similar to that of a swinging pendulum. The simple harmonic curve is a definite improvement over the previous curves. Shock is reduced so that it is no longer serious at moderate speeds.

Construction

The layout (Fig. 2.11) is as follows:

(a) Draw ordinate and abscissa axes, and divide the abscissa into equal parts—usually equal divisions of a circle—4, 6, 8, 10, 12, 16, etc.

(b) Lay off the total rise h on the ordinate. This is the diameter of the harmonic circle.

(c) Draw a semicircle over the rise h, and divide this circle into the same number of equal arcs as chosen in (a).

(d) Project these intercepts to the harmonic-circle diameter, and then to the cam angle divisions. Connect points to obtain the curve.

Characteristics

Referring to Fig. 2.11, we see that the displacement diagram is a cosine curve plotted from points projected from the harmonic circle of radius $h/2$, giving

$$y = (h/2)(1 - \cos \varphi) \quad \text{in.} \tag{2.16}$$

Since the cam rotates β radians while the harmonic circle vector turns through π radians

$$\frac{\varphi}{\pi} = \frac{\theta}{\beta}$$

Solving yields

$$\varphi = \frac{\pi\theta}{\beta} \quad \text{radians} \tag{2.17}$$

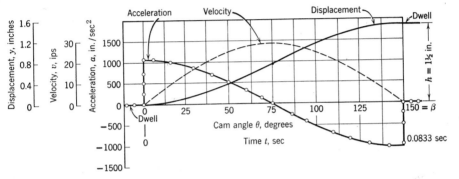

Fig. 2.12. Characteristics of simple harmonic motion curve (D-R-D). (Follower has 1½-in. rise in 150 degrees of cam rotation at 300 rpm.)

where φ = angle of rotation of harmonic-circle vector, radians. Substituting into eq. 2.16 yields the displacement

$$y = \frac{h}{2}\left(1 - \cos \frac{\pi\theta}{\beta}\right) \tag{2.18}$$

Differentiating gives the velocity

$$v = \frac{dy}{dt} = \frac{h\pi\omega}{2\beta} \sin \frac{\pi\theta}{\beta} \quad \text{ips} \tag{2.19}$$

We observe that the velocity curve is a sine curve with points plotted from a rotating vector $h\pi\omega/2\beta$ in length. Differentiating again for the acceleration

$$a = \frac{dv}{dt} = \frac{h}{2}\left(\frac{\pi\omega}{\beta}\right)^2 \cos \frac{\pi\theta}{\beta} \quad \text{in./sec}^2 \tag{2.20}$$

We see that the acceleration is a cosine function with a rotating vector $(h/2)(\pi\omega/\beta)^2$ in length. Note in Fig. 2.12 that the displacement and

velocity curves are smooth and continuous.   However, at the ends
where the dwell meets the simple harmonic curve, there is a sudden
acceleration and discontinuity in the acceleration curve.   This is undesirable for high-speed cams since noise, vibration, and wear result.

Example

Figure 2.12 is plotted with the same data as in the previous articles:
A cam rotates at a constant speed of 300 rpm.   The follower rises 1½
in. with simple harmonic motion in 150 degrees of cam rotation.   Find
the displacement, velocity, and acceleration of the follower after 30
degrees of cam rotation.

Solution

$$\text{Cam angle } \theta = 30 \times \frac{\pi}{180} = 0.524 \text{ radian}$$

$$\text{Cam angular velocity } \omega = 300 \times \frac{2\pi}{60} = 31.42 \text{ rad/sec}$$

$$\text{Total angle } \beta = 150 \times \frac{\pi}{180} = 2.62 \text{ radians}$$

From eq. 2.18, the displacement

$$y = \frac{h}{2}\left(1 - \cos\frac{\pi\theta}{\beta}\right)$$

$$= \frac{1\frac{1}{2}}{2}\left(1 - \cos\frac{\pi \times 0.524}{2.62}\right) = 0.143 \text{ in.}$$

Equation 2.19 gives the velocity

$$v = \frac{h\pi\omega}{2\beta}\sin\frac{\pi\theta}{\beta}$$

$$= \frac{1\frac{1}{2} \times \pi \times 31.42}{2 \times 2.62}\sin\frac{\pi \times 0.524}{2.62} = 16.6 \text{ ips}$$

Last, the acceleration is given by eq. 2.20 as

$$a = \frac{h}{2}\left(\frac{\pi\omega}{\beta}\right)^2\cos\frac{\pi\theta}{\beta}$$

$$= \frac{1\frac{1}{2}}{2}\left(\frac{\pi \times 31.42}{2.62}\right)^2\cos\frac{\pi \times 0.524}{2.62} = 862 \text{ in./sec}^2$$

2.10 DOUBLE HARMONIC CURVE (DWELL-RISE-DWELL CAM)

This unsymmetrical curve (Fig. 2.13) is composed of the difference between two harmonic motions, one being one-quarter of the amplitude and twice the frequency of the other. It has the advantages of the simple harmonic curve with almost complete elimination of high shock and vibration at the beginning of the stroke. The rate of acceleration change at the beginning of the stroke is small, giving smooth action at that point. However, this slow start requires a larger cam for a minimum cam curvature. We see that with a total rise of 1 in. the

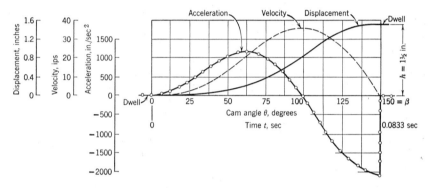

Fig. 2.13.   Characteristics of double harmonic motion curve (D-R-D). (Follower has 1½-in. rise in 150 degrees of cam rotation at 300 rpm.)

follower has risen only 0.02 in. in 20 percent of the lift period, as compared with a rise of 0.15 in. for the simple harmonic curve. The double harmonic curve requires very accurate machining since the errors of cutting its shape (at the start of stroke) usually negate the advantages gained. As a dwell-rise-dwell cam, the limitation of the sudden change in acceleration at the maximum rise point allows only moderate cam speeds.

In general, this curve is better applied to dwell-rise-return-dwell cams (Art. 2.16) in which no sudden acceleration occurs for the complete curve.

The double harmonic curve construction is similar to that of the simple harmonic, consisting of the superposition of the difference of two simple harmonic curves. However, being more complicated it will not be shown, since substituting into the mathematical expressions is easier and more reliable.

Characteristics

The relationship for displacement is

$$y = \frac{h}{2}\left[(1 - \cos \varphi) - \frac{1}{4}(1 - \cos 2\varphi)\right] \text{ in.} \tag{2.21}$$

$$= \frac{h}{2}\left[\left(1 - \cos \frac{\pi\theta}{\beta}\right) - \frac{1}{4}\left(1 - \cos \frac{2\pi\theta}{\beta}\right)\right] \text{ in.} \tag{2.22}$$

The velocity and acceleration by differentiating are

$$v = \frac{dy}{dt} = \frac{h}{2}\frac{\pi\omega}{\beta}\left(\sin \frac{\pi\theta}{\beta} - \frac{1}{2}\sin \frac{2\pi\theta}{\beta}\right) \text{ ips} \tag{2.23}$$

$$a = \frac{dv}{dt} = \frac{h}{2}\left(\frac{\pi\omega}{\beta}\right)^2\left(\cos \frac{\pi\theta}{\beta} - \cos \frac{2\pi\theta}{\beta}\right) \text{ in./sec}^2 \tag{2.24}$$

In Fig. 2.13 are plotted the curves with the same data as in the previous articles: a cam rotates at a constant speed of 300 rpm, the follower rising 1½ in. with double harmonic motion in 150 degrees of cam rotation.

2.11 CYCLOIDAL OR SINE ACCELERATION CURVE (DWELL-RISE-DWELL CAM)

This curve (of the trigonometric family), as the name implies, is basically generated from a cycloid. A cycloid is the locus of a point on a circle which is rolled on a straight line. Applied to cam contours,

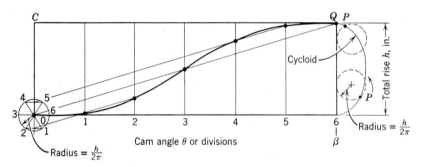

Fig. 2.14. Cycloidal curve construction.

this line is the y axis, the circumference of the circle is made equal to the total rise h, and the radius is equal to $h/2\pi$ (Fig. 2.14). For high speeds, the cycloidal curve is the best of all contours if the accuracy of machining can be maintained at the beginning and the end of the stroke where other curves exhibit their difficulty. It has the lowest vibration, wear, stress, noise, and shock. The reason for its excellent performance is that there is no sudden change in acceleration at the intersection of the dwell periods and the rise curve. It is easy starting, the spring needed to keep the follower on the cam is small, and the side thrust of the translating follower is low.

Construction

The easiest and most accurate method for layout is as follows:

(a) Divide the cam angle into equal parts; in this example, 6 parts were chosen.

(b) A diagonal line $0Q$ is then drawn.

(c) Draw a circle at point 0 of radius equal to $h/2\pi$, which is divided into the same number of equal parts and labeled.

(d) Project these points on the circle to the vertical line $0C$.

(e) Project these points parallel to line $0Q$, giving the intersection to the respective cam angle divisions. Thus, we have points on the curve.

Characteristics

The equation for the displacement is

$$y = \frac{h}{\pi}\left(\varphi - \frac{1}{2}\sin 2\varphi\right) \quad \text{in.} \tag{2.25}$$

$$= \frac{h}{\pi}\left(\frac{\pi\theta}{\beta} - \frac{1}{2}\sin\frac{2\pi\theta}{\beta}\right) \quad \text{in.} \tag{2.26}$$

If we refer to Fig. 2.14, we see that the first term of the equation, $\frac{h}{\pi}\left(\frac{\pi\theta}{\beta}\right)$, is the sloping line $0Q$ and the second term, $\frac{h}{2\pi}\sin\frac{2\pi\theta}{\beta}$, is the harmonic displacement that is subtracted.

Differentiating to find the velocity and acceleration yields

$$v = \frac{dy}{dt} = \frac{h\omega}{\beta}\left(1 - \cos\frac{2\pi\theta}{\beta}\right) \quad \text{ips} \tag{2.27}$$

$$a = \frac{dv}{dt} = \frac{2h\pi\omega^2}{\beta^2}\sin\frac{2\pi\theta}{\beta} \quad \text{in./sec}^2 \tag{2.28}$$

The data is similar to the examples of the previous articles: A cam rotates at a constant speed of 300 rpm. The follower rises $1\frac{1}{2}$ in. with cycloidal motion in 150 degrees of cam rotation. Plotting the results, we obtain Fig. 2.15.

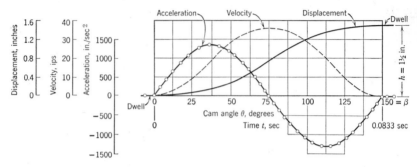

Fig. 2.15.   Characteristics of cycloidal curve (D-R-D).   (Follower has 1½-in. rise in 150 degrees of cam rotation at 300 rpm.)

2.12   PARABOLIC, CONSTANT ACCELERATION, OR GRAVITY CURVE (DWELL-RISE-DWELL CAM)

This curve of the polynomial family has the property of constant positive and negative accelerations.   No other curve will produce a given motion from rest to rest in a given time with so small a maximum acceleration.   This is probably the reason that the curve is erroneously known as the best curve.   It is in many ways the worst of all curves. With perfectly rigid members having no backlash or clearance in the system, the constant acceleration curve would give excellent perform-ance.   However, all members are somewhat elastic and clearance or backlash always exists, especially in the positive drive-roller groove-type follower.   The curve's abrupt change of acceleration at the dwell ends and the transition point produces noise, vibration, and wear, and requires a large spring size.   Thus the parabolic curve should be used only at moderate or lower speeds.   One of the reasons for the popularity of this curve is the ease of determining the inertia forces, which are proportional to the constant accelerations.

Construction

For a symmetrical curve, the layout (Fig. 2.16) is as follows:

(*a*) Plot the ordinate and the abscissa axes with the total rise of the follower as *AB*.

(*b*) Divide the abscissa into equal parts, usually 4, 6, 8, 10, 12, 16, or more.

(*c*) Divide any line *AC* (at any angle to the ordinate) into odd increments of 1, 3, 5, 7, etc., to the midpoint, continuing to the end in reverse order.   In the example, since the abscissa is in 6 equal parts, *AC* had 1, 3, 5, 5, 3, 1 increments.   The 1, 3, 5 represents the positive acceleration period, and the 5, 3, 1 is the negative acceleration period.

The justification of these divisions follows from the parabolic relationship in which the rise at any point is a function of the square of the time or cam angle. See Table 2.1.

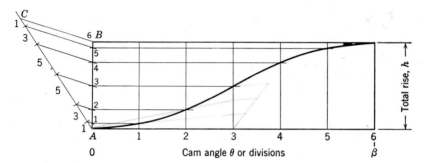

Fig. 2.16. Parabolic curve construction.

Table 2.1

Time t	Displacement Proportional to t^2	Increments of Displacement
1	1	
		3
2	4	
		5
3	9	
		7
4	16	
		9
5	25	
		etc.

(d) Draw a line from the last point C to point B and then other divisions parallel. This now divides the total rise h into its increments. Project these points horizontally to their respective cam divisions giving the curve.

Characteristics

The general displacement for a continuous constant acceleration is

$$ y = v_0 t + \tfrac{1}{2} a t^2 \quad \text{in.} \tag{2.29} $$

where v_0 = initial velocity of the follower, ips. Substituting from eq. 2.3, $t = \theta/\omega$ gives the displacement

$$ y = v_0 \frac{\theta}{\omega} + \frac{1}{2} a \left(\frac{\theta}{\omega}\right)^2 \tag{2.30} $$

Differentiating to give velocity,

$$v = \frac{dy}{dt} = v_0 + at \qquad (2.31)$$

$$= v_0 + a\,\frac{\theta}{\omega} \qquad (2.32)$$

Often the boundary condition is that at the initial point $\theta = 0$, the initial velocity $v_0 = 0$ giving

$$y = \tfrac{1}{2}at^2 \qquad (2.33)$$

$$y = \frac{1}{2}\,a\left(\frac{\theta}{\omega}\right)^2 \qquad (2.34)$$

Let us apply these relationships to establish a basic cam curve. A symmetrical positive and negative acceleration curve will be established

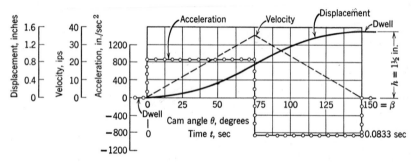

Fig. 2.17. Characteristics of parabolic curve (D-R-D). (Follower has 1½-in. rise in 150 degrees of cam rotation at 300 rpm.)

with a discontinuity at the midpoint of the stroke (Fig. 2.17). In the following, we shall find the characteristics in terms of the cam angle θ, total rise h, and overall angle β. It should be remembered that all values are to be measured from the initial rise point, where cam angle θ is zero.

Cam angle θ from zero to $\beta/2$

The range of the curve is from zero to the maximum velocity, the latter occurring at the transition point. At $\theta = 0$, $v_0 = 0$, and we know that, at the transition point, $\theta = \beta/2$ and $y = h/2$.

Substituting in eq. 2.34 gives

$$\frac{h}{2} = \frac{1}{2}\,a\left(\frac{\beta/2}{\omega}\right)^2 \qquad (2.35)$$

Dividing eq. 2.34 by eq. 2.35 gives the displacement

$$y = 2h \left(\frac{\theta}{\beta}\right)^2 \quad \text{in.} \tag{2.36}$$

For the velocity and acceleration, we differentiate, yielding

$$v = \frac{dy}{dt} = \frac{4h\theta\omega}{\beta^2} \quad \text{ips} \tag{2.37}$$

$$a = \frac{dv}{dt} = \frac{4h\omega^2}{\beta^2} = \text{a constant} \quad \text{in./sec}^2 \tag{2.38}$$

Cam angle θ from $\beta/2$ to β

For the range of the curve from the transition point to the total rise, we have the displacement

$$y_1 = v_m \frac{\theta_1}{\omega} + \frac{1}{2} a \left(\frac{\theta_1}{\omega}\right)^2 \tag{2.39}$$

where y_1 and θ_1 are measured from the transition point and v_m is the maximum velocity (at $\theta = \beta/2$), which from eq. 2.37 is

$$v_m = \frac{4h(\beta/2)\omega}{\beta^2} = \frac{2h\omega}{\beta} \tag{2.40}$$

Also, the acceleration from eq. 2.38 is

$$a = -\frac{4h\omega^2}{\beta^2} \tag{2.41}$$

Substituting in eq. 2.39 gives the displacement

$$y_1 = \frac{2h\theta_1}{\beta}\left(1 - \frac{\theta_1}{\beta}\right) \tag{2.42}$$

However, we know that

$$\theta_1 = \theta - \frac{\beta}{2}$$

and

$$y_1 = y - \frac{h}{2} \tag{2.43}$$

The result of substituting in eq. 2.42 is the displacement, measured from the initial rise,

$$y = h\left[1 - 2\left(1 - \frac{\theta}{\beta}\right)^2\right] \quad \text{in.} \tag{2.44}$$

Differentiating, we find the velocity

$$v = \frac{dy}{dt} = \frac{4h\omega}{\beta}\left(1 - \frac{\theta}{\beta}\right) \quad \text{ips} \tag{2.45}$$

The data is the same as in the previous articles, plotted in Fig. 2.17: A cam rotates at a constant speed of 300 rpm, the follower rising 1½ in. with parabolic motion in 150 degrees of cam rotation.

2.13 CUBIC OR CONSTANT PULSE NO. 1 CURVE (DWELL-RISE-DWELL CAM)

This curve of the polynomial family has a triangular acceleration curve. It is a modification of the parabolic curve, eliminating the abrupt change in acceleration at the beginning and the end of the stroke. This has the advantage of reducing the vibrations, shock, wear, and noise that occur at these points with the parabolic curve. Nevertheless, the cubic curve has the same poor infinite slope characteristic at the midpoint of the acceleration curve, as has the parabolic curve. It has the additional disadvantages of high maximum acceleration and large velocities, necessitating large cams and critical machining. Therefore, this curve is not very practical except when combined with others.

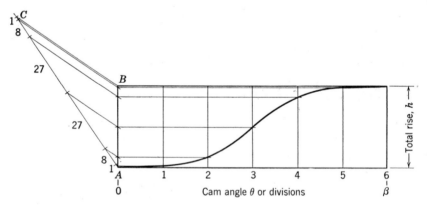

Fig. 2.18. Cubic curve no. 1 construction.

Construction

As the name implies, the displacement is a cubic relationship, with respect to time or the cam angle.

The layout (Fig. 2.18) for a symmetrical curve is as follows:

(a) Plot ordinate and abscissa axes with the total rise of follower as AB.

(b) Divide abscissa into equal parts, usually 4, 6, 8, 10, 12, 16, or more. In the example, 6 parts are chosen.

(c) Divide any line AC at any angle to the ordinate into a cubic relationship. In the example, 1, 8, 27, 27, 8, 1; i.e., 6 parts.

(d) Draw a line from the last point A to point B, and draw other divisions parallel. This now divides the total rise into cubic increments.

(e) Project these points horizontally to their respective cam divisions, and draw the curve.

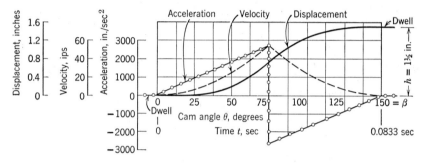

Fig. 2.19. Characteristics of cubic curve no. 1 (D-R-D). (Follower has 1½-in. rise in 150 degrees of cam rotation at 300 rpm.)

Characteristics

By methods shown for the parabolic curve analysis we can solve the constant pulse or cubic curve. We recognize (Fig. 2.19) that the curve is discontinuous and must be considered in halves of the symmetrical curve. The following equations are found:

Cam angle θ from zero to β/2

$$\text{Displacement} \quad y = 4h \left(\frac{\theta}{\beta}\right)^3 \quad \text{in.} \tag{2.46}$$

$$\text{Velocity} \quad v = \frac{12h\omega}{\beta} \left(\frac{\theta}{\beta}\right)^2 \quad \text{ips} \tag{2.47}$$

$$\text{Acceleration} \quad a = \frac{24h\omega^2}{\beta^2} \left(\frac{\theta}{\beta}\right) \quad \text{in./sec}^2 \tag{2.48}$$

$$\text{Pulse} \quad p = \frac{24h\omega^3}{\beta^3} \quad \text{in./sec}^3 \tag{2.49}$$

$$= \text{constant (hence name)}$$

Cam angle θ from $\beta/2$ to β

$$\text{Displacement} \quad y = h\left[1 - 4\left(1 - \frac{\theta}{\beta}\right)^3\right] \qquad (2.50)$$

$$\text{Velocity} \qquad v = \frac{12h\omega}{\beta}\left(1 - \frac{\theta}{\beta}\right)^2 \qquad (2.51)$$

$$\text{Acceleration } a \quad = -\frac{24h\omega^2}{\beta^2}\left(1 - \frac{\theta}{\beta}\right) \qquad (2.52)$$

$$\text{Pulse} \qquad p = \frac{24h\omega^3}{\beta^3} = \text{constant} \qquad (2.53)$$

In Fig. 2.19, we plot the characteristics of the curve with the same data as in the previous articles.

2.14   CUBIC OR CONSTANT PULSE NO. 2 CURVE (DWELL-RISE-DWELL CAM)

This curve is similar to the constant acceleration and the constant pulse no. 1 curves. It differs from these, however, in that there is no abrupt change in acceleration at the transition point and also that its acceleration is a continuous curve for the complete rise. Similar to the constant acceleration curve, it has the disadvantages of abrupt change in acceleration at the beginning and the end of the stroke. This cubic curve has characteristics similar to that of the simple harmonic motion curve. It is not often employed but has advantages when used in combination with other curves. No simple construction method is available.

The characteristic formulas can be found by the same method as that shown for the parabolic curve. They are:

$$\text{Displacement} \quad y = h\frac{\theta^2}{\beta^2}\left(3 - \frac{2\theta}{\beta}\right) \quad \text{in.} \qquad (2.54)$$

$$\text{Velocity} \qquad v = \frac{6h\omega\theta}{\beta^2}\left(1 - \frac{\theta}{\beta}\right) \quad \text{ips} \qquad (2.55)$$

$$\text{Acceleration} \quad a = \frac{6h\omega^2}{\beta^2}\left(1 - \frac{2\theta}{\beta}\right) \quad \text{in./sec}^2 \qquad (2.56)$$

$$\text{Pulse} \qquad p = -\frac{12h\omega^3}{\beta^3} \qquad \text{in./sec}^3 \qquad (2.57)$$

$$= \text{constant (hence name)}$$

Again, the data used to plot Fig. 2.20 is the same as in the previous articles.

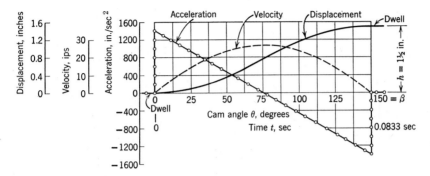

Fig. 2.20. Characteristics of cubic curve no. 2 (D-R-D). (Follower has 1½-in. rise in 150 degrees of cam rotation at 300 rpm.)

2.15 COMPARISON OF BASIC CURVES (DWELL-RISE-DWELL CAM)

Comparing the curves most often used, we shall verify in later chapters that the trigonometric ones (simple harmonic motion, cycloidal, and double harmonic curves) give better overall performance than the basic polynomial family (straight line, parabolic, and cubic curves). The advantages are: smaller cams, lower follower side thrust, cheaper manufacturing cost, easier layout, and easier duplication. For the mathematical summary of the equations, the reader is referred to Table 2.2.

Table 2.3 is a compilation and comparison of the basic curves, showing the cam size ratio, velocity, acceleration, comments, and speed application. The cam size ratio gives the reader a comparison of cams having the same maximum pressure angle, using the straight-line curve as a unit size. In Fig. 2.21, we have the curves of the previous examples superimposed for quick reference. Now let us summarize the application of all curves.

The *straight-line curve* is poor at any speed due to the sudden contour "bump." It is not suggested for use with dwells at either end.

The *straight-line circular arc curve* reduces the shock of the straight line; still its application is not suggested except at low speeds.

The *circular arc curve* has the advantage of further reducing the shock but at the expense of a large pressure angle for the same size cam or a large cam for the same pressure angle. It is suggested for low speeds only, if at all.

The *elliptical curve,* if properly proportioned, has characteristics that can compare with the simple harmonic motion curve. Construction difficulty and large cam size do not justify its preference over

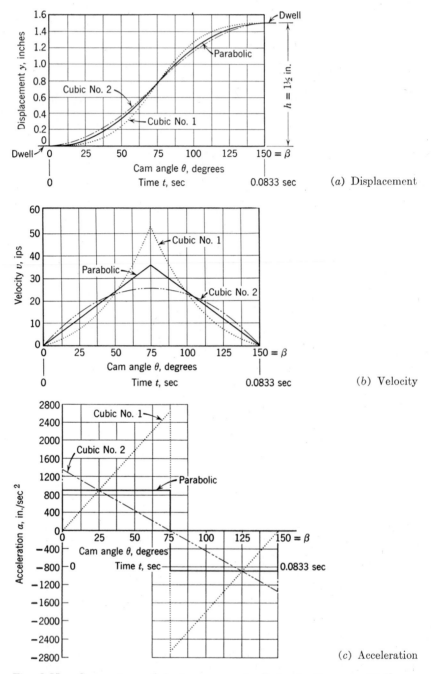

(a) Displacement

(b) Velocity

(c) Acceleration

Fig. 2.21.   Comparison of basic curves—dwell-rise-dwell cam.   (Follower

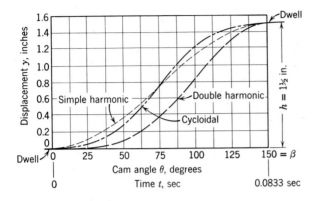

(a) Displacement

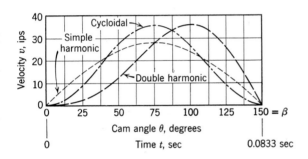

(b) Velocity

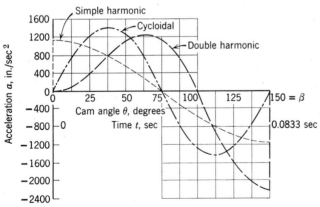

(c) Acceleration

has 1½-in. rise in 150 degrees of cam rotation at 300 rpm.)

Table. 2.2   Characteristic Equations of Basic Curves

Curves		Displacement y, in.	Velocity v, ips	Acceleration a, in./sec²
Straight line		$\dfrac{h\theta}{\beta}$	$\dfrac{\omega h}{\beta}$	0
Circular arc		$H - [H^2 - (R_p\theta)^2]^{1/2}$	$\dfrac{\omega R_p^2\theta}{[H^2 - (R_p\theta)^2]^{1/2}}$	$\dfrac{(\omega R_p H)^2}{[H^2 - (R_p\theta)^2]^{3/2}}$
Simple harmonic motion (SHM)		$\dfrac{h}{2}\left(1 - \cos\dfrac{\pi\theta}{\beta}\right)$	$\dfrac{h\pi\omega}{2\beta}\sin\dfrac{\pi\theta}{\beta}$	$\dfrac{h}{2}\left(\dfrac{\pi\omega}{\beta}\right)^2\cos\dfrac{\pi\theta}{\beta}$
Double harmonic		$\dfrac{h}{2}\left[\left(1 - \cos\dfrac{\pi\theta}{\beta}\right) - \dfrac{1}{4}\left(1 - \cos\dfrac{2\pi\theta}{\beta}\right)\right]$	$\dfrac{h\pi\omega}{2\beta}\left(\sin\dfrac{\pi\theta}{\beta} - \dfrac{1}{2}\sin\dfrac{2\pi\theta}{\beta}\right)$	$\dfrac{h}{2}\left(\dfrac{\pi\omega}{\beta}\right)^2\left(\cos\dfrac{\pi\theta}{\beta} - \cos\dfrac{2\pi\theta}{\beta}\right)$
Cycloidal		$\dfrac{h}{\pi}\left(\dfrac{\pi\theta}{\beta} - \dfrac{1}{2}\sin\dfrac{2\pi\theta}{\beta}\right)$	$\dfrac{h\omega}{\beta}\left(1 - \cos\dfrac{2\pi\theta}{\beta}\right)$	$\dfrac{2h\pi\omega^2}{\beta^2}\sin\dfrac{2\pi\theta}{\beta}$
Parabolic or constant acceleration	$\dfrac{\theta}{\beta} \leqq 0.5$	$2h\left(\dfrac{\theta}{\beta}\right)^2$	$\dfrac{4h\omega\theta}{\beta^2}$	$4h\dfrac{\omega^2}{\beta^2}$
	$\dfrac{\theta}{\beta} \geqq 0.5$	$h\left[1 - 2\left(1 - \dfrac{\theta}{\beta}\right)^2\right]$	$\dfrac{4h\omega}{\beta}\left(1 - \dfrac{\theta}{\beta}\right)$	$-4h\dfrac{\omega^2}{\beta^2}$
Cubic no. 1 or constant pulse no. 1	$\dfrac{\theta}{\beta} \leqq 0.5$	$4h\left(\dfrac{\theta}{\beta}\right)^3$	$\dfrac{12h\omega}{\beta}\left(\dfrac{\theta}{\beta}\right)^2$	$\dfrac{24h\omega^2}{\beta^2}\left(\dfrac{\theta}{\beta}\right)$
	$\dfrac{\theta}{\beta} \geqq 0.5$	$h\left[1 - 4\left(1 - \dfrac{\theta}{\beta}\right)^3\right]$	$\dfrac{12h\omega}{\beta}\left(1 - \dfrac{\theta}{\beta}\right)^2$	$-\dfrac{24h\omega^2}{\beta^2}\left(1 - \dfrac{\theta}{\beta}\right)$
Cubic no. 2 or constant pulse no. 2		$h\left(\dfrac{\theta}{\beta}\right)^2\left(3 - 2\dfrac{\theta}{\beta}\right)$	$\dfrac{6h\omega\theta}{\beta^2}\left(1 - \dfrac{\theta}{\beta}\right)$	$\dfrac{6h\omega^2}{\beta^2}\left(1 - 2\dfrac{\theta}{\beta}\right)$

where h = maximum rise of follower, in.
β = cam angle of rotation to give rise h, radians.
ω = cam angular velocity, rad/sec.
θ = cam angle rotation for follower displacement y, radians.
H = radius of the circular arc, in.
R_p = radius of the pitch circle, in.

others. It can be employed for moderate speeds with an 11 : 8 major-to-minor axis ratio.

The *simple harmonic curve* is easy to construct and calculate and also gives reasonable cam-follower action at moderate speeds. The sudden change in acceleration at the beginning and end of the action in dwell-rise-dwell curves precludes high-speed application. However, this curve has low follower side thrust, smooth starting, and a reasonable follower spring size compared to the basic polynomial curves.

The *double harmonic curve* is similar to the simple harmonic curve with the additional advantage of smoother action at the start. However, this is achieved at the expense of a larger cam for the same minimum cam curvature. Also, cutting at the initial points of the curve requires extreme accuracy to gain the advantages of the mathematical curve. The double harmonic curve is utilized to best advantage in a dwell-rise-return-dwell cam. In this application, lower vibration, favorable wear, easy start, low follower side thrust, and small spring size result. In general, sudden acceleration at the maximum rise point precludes its choice as a dwell-rise-dwell cam in lieu of other simpler ones.

The *cycloidal curve,* for most machine requirements, is the best of all. Since it has no abrupt change in acceleration, it gives the lowest vibrations, wear, stress, noise, and shock. It is easy starting, requires small springs, and induces low follower side thrust. However, the necessary accuracy of fabrication is higher than for low-speed curves.

The *parabolic curve,* with perfectly rigid members and no backlash or clearance in the linkages, would be excellent. But, since these conditions do not occur, whether a compression spring or positive drive follower is employed to maintain constraint, the curve gives poor performance at high speeds. Its use should be limited to moderate speeds or lower. It is inferior in wear, pressure angle, spring size, etc., as compared to the trigonometric curves.

The *cubic curve no. 1* is not practical due to large cam sizes. In addition, generally high accuracy of cam cutting is necessary to gain the advantages of the mathematical curve. This will offer difficulty.

The *cubic curve no. 2* has characteristics similar to the simple harmonic curve.

2.16 APPLICATION OF BASIC CURVES TO DWELL-RISE-RETURN-DWELL CAM

In this article we shall discuss the application of the symmetrical basic curves to dwell-rise-return-dwell (D-R-R-D) cams. For un-

Table. 2.3   Practical Comparison of Basic Curves for Dwell-Rise-Dwell Cams

Basic Curves	Cam Size Ratio	Follower Motion		Comments	Speed Application
		Velocity	Acceleration		
Straight line	1	Constant	Infinite at ends	Large shock or "bump" at ends of curve	Impractical with dwell ends
Straight-line circular arc	—	Less at ends than the above	Large at ends	The shock is less than the above but is still serious	Low speeds if at all
Circular arc	—				
Elliptical		Values depend on major-to-minor axis ratio			
Simple harmonic motion (SHM)	1.6	Changes from zero to maximum (at mid-point) to zero again	Further reduced at ends — zero at midstroke	Further reduction in shock which is not serious at moderate speeds. Low translating follower jamming and small spring size result	Fair performance at moderate speeds
Cycloidal	2	Slower at start and stop than above but greater at midpoint	Smooth application	Best of all simple curves. Low vibration, stresses, noise, and wear. High accuracy needed in fabricating. Low translating follower jamming and small spring size result	High speeds

Parabolic or constant acceleration	2	Uniform increase to mid-stroke, then uniform decrease	Least maximum value of all. But abrupt application of acceleration	Beware of backlash at midpoint (groove cams). This gives high shock, noise, and wear	Low to moderate speeds
Cubic no. 1	3	—	Abrupt acceleration at midpoint	Pressure angle, cam size, and follower velocity are too large. Manufacturing accuracy is difficult to obtain	Low speeds if at all
Cubic no. 2	1½		Similar to the simple harmonic motion curve		
Double harmonic	2	Slowest beginning of all	Smoothest application of initial rise	Smooth application of initial load. Gives low translating follower jamming and small spring size. Requires highest accuracy of all curves. It is best as a D-R-R-D cam	Moderate to high speeds

symmetrical basic curves, the reader is referred to Chapter 6. Much
of the previous information on the use of basic curves in dwell-rise-
dwell cam applies to this D-R-R-D cam with certain slight changes.

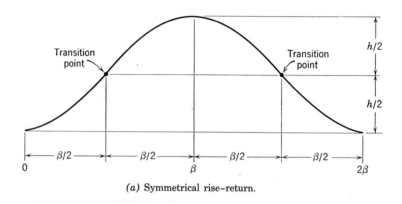

(a) Symmetrical rise–return.

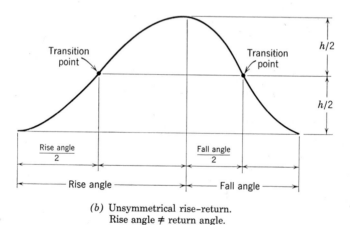

(b) Unsymmetrical rise–return.
Rise angle ≠ return angle.

**Fig. 2.22.   Displacement   diagrams.   Dwell-rise-return-dwell   cam   having
symmetrical basic curves.**

Dwell-rise-return-dwell cams, having the symmetrical basic curves
of this chapter, may be classified:

(1) Symmetrical rise and fall displacement curve, Fig. 2.22a. In
this figure we see that both the rise and the fall take place in the same
time or through the same cam angle.

(2) Unsymmetrical rise and fall displacement curve (Fig. 2.22*b*), i.e., the cam angle of rise is different from the cam angle of fall.

Dwell-rise-return-dwell cams—completely symmetrical

Let us now plot some basic curves for a dwell-rise-return-dwell cam. In Fig. 2.23*a*, for comparison we see the symmetrical parabolic, simple

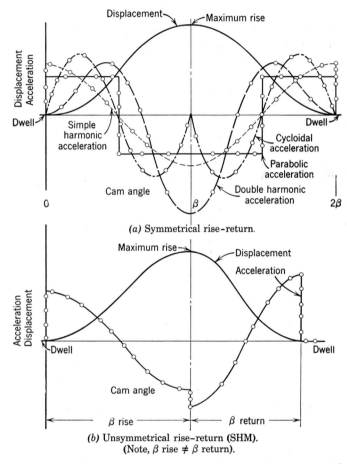

(*a*) Symmetrical rise-return.

(*b*) Unsymmetrical rise–return (SHM).
(Note, β rise $\neq$ β return).

Fig. 2.23. Dwell-rise-return-dwell cam application of symmetrical basic curves.

harmonic motion, cycloidal and double harmonic curves. We know that the displacement and velocity curves of all are smooth and continuous between the dwell ends. However, the parabolic and simple harmonic acceleration curves exhibit the same difficulty as those in the dwell-rise-dwell application, i.e., sudden change at the ends. This shortcoming is known to produce inferior performance under high speeds. The cycloidal curve, on the other hand, which is one of the best dwell-rise-dwell curves, has an abrupt change in acceleration at the maximum rise point. This is not desirable, since vibration and difficult machining result. The magnitude of these factors, of course, depends on cam speed and flexibility of the parts. However, we note that at the maximum rise point neither the parabolic nor the simple harmonic have any abrupt change. We also see that the best high-speed curve, from the standpoint of slow starting and stopping action and no abrupt change in acceleration, is the double harmonic curve. If it can be manufactured with the accuracy necessary, it is suggested as the best basic curve for high-speed action.

Dwell-rise-return-dwell cam—unsymmetrical rise return

In this type of cam, we may observe another difficulty at the point of maximum rise, Fig. 2.23b. Using the simple harmonic motion curve as an example, we plot the action with the rise occurring in a shorter time or cam angle than the return. At the peak rise point, we observe an undesirable abrupt change in acceleration, with vibrations, noise, and wear resulting. This abrupt acceleration may be ignored at moderate or lower cam speeds.

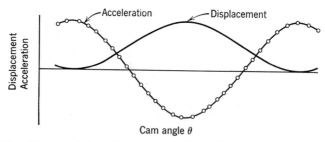

Fig. 2.24. **Characteristics of symmetrical simple harmonic rise-return-rise cam. (Note continuous acceleration curve.)**

2.17 APPLICATION OF BASIC CURVES TO RISE-RETURN-RISE CAM

The rise-return-rise cam has no dwells. Therefore, a complete symmetrical and smooth acceleration characteristic would be acceptable. The simplest method for giving this symmetrical action is by using a crank or eccentric rather than a cam mechanism. The displacement, velocity, and acceleration curves have a continuous sine or cosine curve characteristic (Fig. 2.24). An eccentric is more compact and cheaper to fabricate than a cam. Thus, a cam is generally not needed for this R-R-R action. However, if an asymmetrical action is necessary, the simple harmonic motion cam curve should be employed.

References

1. J. R. Longstreet, "Systematic Correlation of Motion," *Mach., Des. 25,* p. 215 (Dec. 1950).
2. W. Richards, "Cam Design, Constant Velocity Motion Cam," *Mach., London 48,* p. 473 (July 16, 1936).
3. W. Richards, "Cam Design, the Design of Uniformly Accelerated and Retarded Motion Cam," *Mach., London 52,* p. 573 (Aug. 11, 1938).
4. W. Richards, "Cam Design, the Design of Uniformly Increasing Acceleration and Retardation Cam," *Mach., London 53,* p. 365 (Dec. 22, 1938).
5. W. Richards, "High Speed Cam Design," *Mach., London 54,* p. 621 (Aug. 17, 1939) and *55,* p. 105 (Oct. 26, 1939).
6. W. Richards, "The Harmonic Motion Cam," *Mach., London, 55,* p. 481 (Feb. 1, 1940).

CHAPTER THREE

Cam Size Determination

3.1

In this chapter we shall present direct mathematical means for determining the cam size for proper follower action. The minimum cam size is basically affected by (1) the pressure angle or the steepness of its profile, (2) the curvature or sharpness of the profile, and (3) the size of the cam shaft. Minimum size is desirable because of space limitations, unbalance at high speeds, longer paths of follower movement, and correspondingly higher wear.

This chapter is intended as an introduction to Chapter 4, Cam Profile Determination. Some engineers prefer to circumvent this presentation temporarily and go directly to the next chapter. They utilize the empirical method of constructing the cam, observing the pressure angle and cam curvature, and repeating the layout until the proper size has been obtained. This trial-and-error procedure is, of course, more time consuming.

Pressure Angle

3.2 INTRODUCTION

No other single factor in cam design is more often mentioned and less understood than the pressure angle. In these articles we shall endeavor to correct this inadequacy. Visual inspection of a layout or an arbitrary limit for the maximum pressure angle may prove safe in design. However, a fundamental understanding and use of the factors that determine the pressure angle, in addition to experience and proper analysis of the problem, are necessary. We shall show that the primary emphasis on the pressure angle is the distribution of forces in the follower.

Thus, it is the maximum value that is of concern. The maximum pressure angle establishes the cam size, torque, loads, accelerations, wear life, and other pertinent factors.

3.3 RADIAL CAM LAYOUT DISTORTION

If the pitch curve of the displacement diagram is plotted on a radial cam, we see that it is distorted towards the cam center. That is, the displacement-diagram pressure angle is not the same as the pressure angle measured on the actual radial cam. As an example, in Fig. 3.1a, let us take a straight-line pitch curve OB in the displacement diagram.

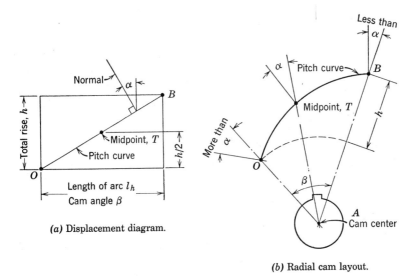

(a) Displacement diagram.

(b) Radial cam layout.

Fig. 3.1. Radial cam layout distortion. [Pressure angles different in (a) from those in (b).]

This pitch curve has a constant pressure angle α for a total rise h in β cam degrees. Let point T be the midpoint of rise OB, and, in Fig. 3.1b, let A be the cam center. By trial and error, choose a radius AT, so that the pressure angle of the pitch curve at point T is equal to α. Observe that, whereas we had a constant slope in Fig. 3.1a, the pressure angle is distorted when laid out on the radial cam (Fig. 3.1b). It is larger than α below point T and smaller than α above point T. Therefore, the closer the pitch curve is to center of cam, the larger the pressure angle. In other words, the pressure angle can be reduced by using a larger radius and thus a larger cam. It should be remembered that radial distortion affects any pitch curve in a manner similar to this example.

3.4   MAXIMUM PRESSURE ANGLE VS. SIDE THRUST— RADIAL CAM, TRANSLATING ROLLER FOLLOWER

Let us now analyze the side thrust due to excessive pressure angle on a translating radial roller follower. In this article, we shall show that the pressure angle should be limited in accordance with the length of follower overhang, its guide bearing length, the coefficient of friction, and the rigidity of the follower stem. First, assume that the friction in the follower roller bearing is negligible. Let

μ = coefficient of friction between follower stem and its guide bearing.

N_1 and N_2 = forces normal to follower stem, lb.

F = force normal to cam profile, lb.

L_o = total external load on follower (includes weight, spring, force, inertia, friction etc.), lb.

A = follower overhang, in.

B = follower bearing length, in.

α = pressure angle, degrees.

α_m = maximum pressure angle, degrees.

W = radius of follower stem, in.

Figure 3.2 shows the direction of cam rotation, the normal forces on the follower, and the frictional forces opposing the motion of the follower. For static equilibrium, the sum of forces along the vertical axis is

$$\Sigma F_y = 0 = -L_o + F \cos \alpha - \mu N_1 - \mu N_2 \qquad (3.1)$$

Let points p and q be the intersection of N_1 and N_2 on the line of follower motion. From statics, the sum of the moments is

$$\Sigma M_p = 0 = -FA \sin \alpha + N_1 B - \mu N_1 W + \mu N_2 W \qquad (3.2)$$

$$\Sigma M_q = 0 = -F(A + B) \sin \alpha + N_2 B - \mu N_1 W + \mu N_2 W \qquad (3.3)$$

Simplifying eq. 3.2 and eq. 3.3 and assuming $\mu N_1 W$ and $\mu N_2 W$ equal zero since they are negligible yields

$$N_1 = \frac{A}{B} F \sin \alpha \qquad (3.4)$$

$$N_2 = \frac{A + B}{B} F \sin \alpha \qquad (3.5)$$

Substituting eq. 3.4 in eq. 3.1 and eliminating N_1

$$0 = -L_o + F \cos \alpha - \mu F \frac{A}{B} \sin \alpha - \mu N_2 \qquad (3.6)$$

Substituting eq. 3.5 in eq. 3.6 to eliminate N_2 gives the external load

$$L_o = F \left[\cos \alpha - \mu \left(\frac{2A + B}{B} \right) \sin \alpha \right]$$

Solving for the force normal to the cam gives

$$F = \frac{L_o}{\cos \alpha - \mu \left(\dfrac{2A + B}{B} \right) \sin \alpha} \quad (3.7)$$

The normal force F is a maximum (equals infinity) which means that the follower will jam in its guide when the denominator of eq. 3.7 equals zero. Therefore,

$$\cos \alpha_m - \mu \left(\frac{2A + B}{B} \right) \sin \alpha_m = 0$$

The maximum pressure angle for locking the follower in its guide

$$\alpha_m = \tan^{-1} \frac{B}{\mu(2A + B)} \quad (3.8)$$

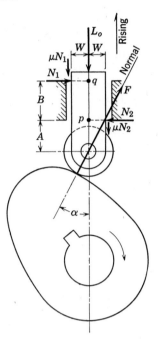

Fig. 3.2. Radial cam translating roller follower force distribution.

Let us substitute some trial values to compare the magnitude of the results. If we let $A = B$ and assume the values for the coefficient of friction of bronze on steel to be:

$$\mu(\text{kinetic}) = 0.10 \text{ and } \mu(\text{static}) = 0.15$$

Substituting in eq. 3.8, we find the maximum pressure angle for each condition:

$$\alpha_m = \tan^{-1} \frac{B}{(0.10)(2B + B)} = 73\tfrac{1}{2} \text{ degrees} \quad \text{for } \mu = 0.10$$

$$\alpha_m = \tan^{-1} \frac{B}{(0.15)(2B + B)} = 66 \text{ degrees} \quad \text{for } \mu = 0.15$$

Note that these values and the derivation of eq. 3.8 are based on the ideal assumption that the follower is perfectly rigid. Thus, the coefficient of friction may actually reach a value of 0.25 or more, depending on flexibility and backlash of the follower. A flexible stem may "dig" into the lower corner of the bearing. Therefore, the suggested guide in practice is to keep the coefficient of friction μ, the follower overhang

A, and the backlash as small as possible with the bearing length B as large as possible. In addition, the follower stem should be made rigid. Fulfilling these requirements will give the largest pressure angle and the smoothest follower action. Generally the safe limiting pressure angle in practice is 30 degrees. However, for light loads with accurate low-friction bearings, the author was successful using a pressure angle as high as $47\frac{1}{2}$ degrees. Note that commercially available ball bushings for the linear moving stem have provided low friction and little backlash.

We may observe that the follower jamming is only of concern when the follower moves in a direction opposite that of the external load L_o. As shown in Fig. 3.2, jamming occurs during the rise period only. During the fall period, the size of the maximum pressure angle is generally not limited in proper cam design. However, the author has seen machine installations in which the follower had driven the cam during the fall action. It occurred with a chain-driven cam and a spring-loaded follower. The spring force acting on an excessive pressure angle of fall, produced detrimental shock and fluctuating action in absorbing the backlash in the system.

In changing from rise to fall, the reactions N_1 and N_2 will reverse their directions to the other side of the follower stem. This necessitates that the clearance between the follower and its bearing guide should be kept to a minimum to limit the noise and shock.

All of the foregoing discussion refers to an offset follower as well. However, mathematical relationships are not shown because of their complexity. The advantage of offsetting is that the pressure angle may be reduced during that portion of the action (rise or fall) which has the detrimental side thrust. Furthermore, decreasing the pressure angle in one portion increases the pressure angle in the other less important portion of the cam action. Almost complete elimination in pressure-angle side thrust may be accomplished by using a secondary follower (Fig. 4.11). In this design the undesirable force components are prevented from jamming the follower and producing wear. In Art. 3.10, we see a side-thrust comparison of some curves on high-speed cams.

3.5 SIDE THRUST—TRANSLATING FLAT-FACED FOLLOWER

The side thrust or jamming effect on all flat-faced followers is inconsequential compared to that on the roller follower. The translating flat-faced follower is shown (Fig. 3.3) with normal forces N_1 and N_2. It is obvious that their directions change, depending on the distance q. It follows that the pressure angle is zero at all times, allowing the cam to be much smaller. Jamming of the follower is caused by the net effect of the opposing moments of forces F and μF. We have indicated in Art. 3.4, that, for best performance, the ratio B/A should be kept large,

coefficient of friction μ and eccentricity q should be kept small, and the rod rigidity should be as high as possible. The fact that these values are obviously not critical compared with the roller follower is the inherent advantage of the flat-faced follower.

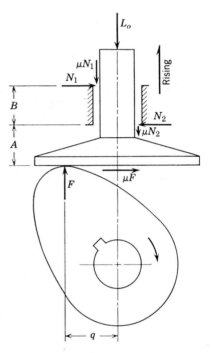

Fig. 3.3. Radial cam translating flat-faced follower force distribution.

3.6 PRESSURE ANGLE VS. FORCE DISTRIBUTION—OSCILLATING ROLLER FOLLOWER

The oscillating roller follower is generally an improvement over the translating type of Art. 3.4 in that the distribution of forces allows a much larger limiting pressure angle for satisfactory performance. Thus, with an oscillating follower, a smaller cam may be used. It is suggested for best performance and highest accuracy that the trace point arc of motion should pass near or through the cam center, Fig. 4.14. Also, by proper location of the follower pivot, it becomes virtually impossible to jam the follower, no matter how steep the cam surface. The extreme limiting condition is to make the pressure angle small enough to prevent the cam normal force from passing through the follower pivot. Locking results with this detrimental force distribution. Often, for ultimate performance, frequent trials must be made by layout to establish the best dimensional proportions under the given design conditions. Therefore, the side thrust will not exist with the properly designed oscillating roller follower.

3.7 PRESSURE ANGLE—RADIAL CAM, TRANSLATING OFFSET ROLLER FOLLOWER

In this article we shall discuss the relationship between the size of cam, pressure angle, and type of curve for the general case of a translating offset roller follower. Let

r = radius to trace point, in.
e = eccentricity or offset, in.
y_0 = vertical displacement of prime circle, in.
y = displacement of follower, in.
R_a = prime circle radius, in.
α = pressure angle, degrees.
$z = y_0 + y$ = vertical distance to trace point, in.
γ = angle between radius r and tangent to pitch curve, degrees.
ω = cam angular velocity, rad/sec.
θ = cam angle of rotation for displacement y, radians.
t = time for cam to rotate angle θ, sec.

The following relationships can be observed in Fig. 3.4. First, the prime circle radius

$$R_a = (y_0{}^2 + e^2)^{\frac{1}{2}} \qquad (3.9)$$

Also, the radius vector from the center of rotation to the trace point at any angle of rotation θ is

$$r = (R_a{}^2 + 2y_0 y + y^2)^{\frac{1}{2}} \qquad (3.10)$$

and

$$\tan \gamma = \frac{r}{dr/d\theta} \qquad (3.11)$$

Solving these equations, it could be shown that the pressure angle for the offset follower is

$$\tan \alpha = \frac{\dfrac{z^2}{\omega} \dfrac{dy}{dt} - er^2}{z\left(\dfrac{e}{\omega} \dfrac{dy}{dt} + r^2\right)} \qquad (3.12)$$

If the offset e equals zero, i.e., for radial followers of radial cams and axial translating followers of cylindrical cams, the pressure angle becomes

$$\tan \alpha = \frac{1}{z} \frac{dy}{d\theta} \qquad (3.13)$$

$$= \frac{1}{z\omega} \frac{dy}{dt} \qquad (3.14)$$

Let us solve a problem to show the effect of offsetting on the pressure angles of a cam.

Example

A radial cam with a prime circle diameter of 4 in. rotates at 300 rpm. The follower rises 1½ in. with SHM in 150 degrees of cam rotation.

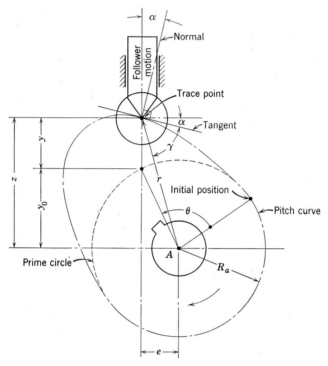

Fig. 3.4. Radial cam translating offset roller follower.

When the cam angle is 30 degrees, find the value of the pressure angle for (a) follower offset opposite the cam rotation equal to ½ in.; (b) radial follower.

Solution

The basic data for the problem was found in Art. 2.9, in which

Cam angle	$\theta =$	0.524 radians
Cam angular velocity	$\omega =$	31.42 radians/sec
Total angle	$\beta =$	2.62 radians
Displacement of follower	$y =$	0.143 in.
Velocity of follower	$dy/dt =$	16.6 ips

(a) From eq. 3.9

$$R_a{}^2 = (y_0{}^2 + e^2)^{\frac{1}{2}}$$

$$(2)^2 = [y_0{}^2 + (\tfrac{1}{2})^2]^{\frac{1}{2}}$$

Therefore the base-circle displacement $y_0 = 1.942$ in. Also, from Fig. 3.4, we see

$$z = y_0 + y$$
$$= 1.942 + 0.143 = 2.085 \text{ in.}$$

Substituting in eq. 3.10, the radius vector is

$$r = (R_a^2 + 2y_0 y + y^2)^{1/2}$$
$$= [2^2 + 2 \times 1.942 \times 0.143 + (0.143)^2]^{1/2}$$
$$= 2.143 \text{ in.}$$

The pressure angle from eq. 3.12

$$\alpha = \tan^{-1} \frac{\dfrac{z^2}{\omega} \dfrac{dy}{dt} - er^2}{z\left(\dfrac{e}{\omega} \dfrac{dy}{dt} + r^2\right)}$$

$$= \tan^{-1} \frac{\dfrac{(2.085)^2}{31.42}(16.6) - \frac{1}{2}(2.143)^2}{(2.085)\left[\dfrac{\frac{1}{2}}{31.42}(16.6) + (2.143)^2\right]} = 0.50 \text{ degrees}$$

(b) For radial cams the eccentricity e equals zero. Therefore, referring to Fig. 3.4, we see

$$r = z = y_0 + y$$
$$= 2 + 0.143 = 2.143 \text{ in.}$$

Solving eq. 3.14 gives the pressure angle

$$\alpha = \tan^{-1} \frac{1}{z\omega} \frac{dy}{dt}$$

$$= \tan^{-1} \frac{16.6}{(2.143)(31.42)} = 13.85 \text{ degrees}$$

3.8 EXACT MAXIMUM PRESSURE ANGLE AND RELATED CAM SIZE—RADIAL CAM, TRANSLATING ROLLER FOLLOWER

In many problems both the maximum pressure angle and the cam angle where it occurs is desired. For the offset roller follower, the solution of eq. 3.12 to find this is too complicated mathematically, requiring trial-and-error methods. Therefore, for both the translating offset and oscillating roller followers, the maximum pressure angle is established most easily by trial-and-error *layout* methods.

For the radial follower, we can derive an exact expression for the maximum pressure angle. Let

$$R_p = \text{radius of pitch circle, in.}$$

$$\theta_p = \text{cam angle at pitch point, radians}$$

Thus, we could illustrate by use of eq. 3.13 that the pitch-circle radius, i.e., at the pitch point, or point of maximum pressure angle, is

$$R_p = \frac{\left(\dfrac{dy}{dt}\right)_p^2}{\left(\dfrac{d^2y}{dt^2}\right)_p} \tag{3.15}$$

The subscript p refers to the pitch point value.

Also, the maximum pressure angle utilizing eq. 3.12 is

$$\tan \alpha_m = \frac{\left(\dfrac{d^2y}{dt^2}\right)_p}{\omega \left(\dfrac{dy}{dt}\right)_p} \tag{3.16}$$

With these equations we can, if they are solvable, construct the smallest radial cam with a limiting pressure angle. However, they have restricted practical value since they are easily solved only for some basic curves. For example, for the simple harmonic motion curve, we find

$$R_p = \frac{h \left(\sin \dfrac{\pi \theta_p}{\beta}\right)^2}{2 \cos \dfrac{\pi \theta_p}{\beta}} \tag{3.17}$$

$$\tan \alpha_m = \frac{\pi}{\beta} \cot \frac{\pi \theta_p}{\beta} \tag{3.18}$$

We know that one of the first steps in cam design is to find the pitch-circle radius R_p and the angle θ_p at which α_m occurs.

Example

A radial cam rotates 180 rpm with the follower rising 4 in. with SHM in 150 degrees of rotation. Find the exact pitch point location and pitch-circle radius for a maximum pressure angle of 30 degrees.

Solution

The total cam angle

$$\beta = 150 \, \frac{\pi}{180} = 2.62 \text{ radians}$$

From eq. 3.18 the maximum pressure angle is

$$\tan \alpha_m = \frac{\pi}{\beta} \cot \frac{\pi\theta_p}{\beta}$$

$$\tan 30 = \frac{\pi}{2.62} \cot \frac{\pi\theta_p}{2.62}$$

Solving, we obtain the pitch point location at cam angle $\theta_p = 53.6$ degrees or 0.935 radian.

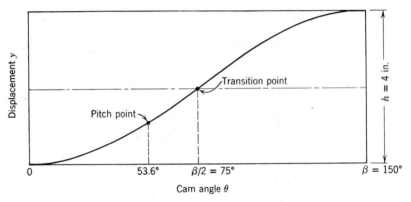

Fig. 3.5. **Displacement diagram of example in Art. 3.6.**

From eq. 3.17, the radius of the pitch circle

$$R_p = \frac{h \left(\sin \frac{\pi\theta_p}{\beta} \right)^2}{2 \cos \frac{\pi\theta_p}{\beta}}$$

$$= \frac{4 \sin^2 \frac{\pi(0.935)}{2.62}}{2 \cos \frac{\pi(0.935)}{2.62}} = 3.77 \text{ in.}$$

This gives the information necessary to construct the cam profile having a maximum pressure angle of *exactly* 30 degrees. In the displacement diagram (Fig. 3.5), we see the location of the pitch point at $\theta_p = 53.6$

degrees and the transition point at $\beta/2 = 75$ degrees. Distortion, discussed in Art. 3.3, causes this difference.

3.9 MAXIMUM PRESSURE ANGLE AND RELATED CAM SIZE—TRANSLATING ROLLER FOLLOWER

In this article we shall establish an approach for calculating the maximum pressure angle for cams in general. It is an *exact* method for *cylindrical cams* and a simplified, *approximate* method for *radial cams*.

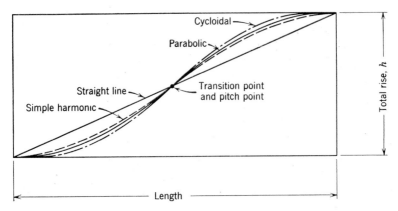

Fig. 3.6. Comparison of some basic curves. (Note: maximum pressure angles are different.)

We know that, contrary to radial cams, the displacement diagram is a true development of the cylindrical cam pitch curve drawn to scale. The cam and displacement diagram slopes are the same.

If we compare, in Fig. 3.6, some of the basic curves, we see for the same total rise and a length of displacement diagram that the maximum slopes or pressure angles are different. Or, if we hold the maximum pressure angle to a limiting value, which is usually done, the length of pitch-circle arc or size of the cam would vary accordingly. For the curves shown in Fig. 3.6, the straight-line curve gives the smallest cam with the cycloidal and parabolic curves the largest. Therefore, the maximum pressure angle depends on two factors—the cam size and the basic cam curve.

Let us establish the mathematical relationship between these parameters. In Fig. 3.7, we observe the displacement diagram with the developed pitch circle passing through the pitch point. The pitch and transition points are at the same place. For most radial cam designs, this assumption is within the accuracy of the pressure angle limitation.

In general, small cams give greatest error in radial cam construction.
Let

l_h = developed length of cam for rise h, in.
f = cam factor for each curve.
h = maximum displacement of follower, in.
β = cam angle rotation for displacement h, radians.

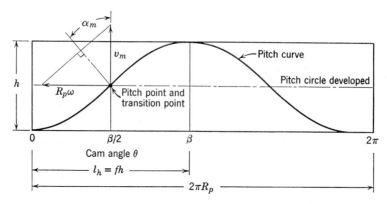

**Fig. 3.7. Displacement diagram. (Note: pitch point and transition point
are considered at the same place.)**

Note that with cylindrical cams the pitch circle refers to the circle
passing through the mean length of roller follower contacting the cam,
see Fig. 3.9.

For any curve (Fig. 3.7) the length of the displacement diagram may
be considered to be

$$l_h = fh$$

But, by proportion since l_h is also the arc of the pitch circle

$$l_h = R_p \beta$$

Therefore, the pitch-circle radius is

$$R_p = \frac{fh}{\beta} \tag{3.19}$$

Cam factor f

Let us indicate the method for finding these factors, using the simple
harmonic motion curve as an example. We know the maximum pressure angle is assumed at the pitch point (Fig. 3.7). This maximum

pressure angle is

$$\tan \alpha_m = \frac{v_m}{R_p \omega} \qquad (3.20)$$

where $v_m = (dy/dt)_{\max}$ = maximum follower velocity, ips. However, from eq. 2.19, the maximum velocity of the follower for simple harmonic motion at a cam angle θ equal to $\beta/2$ is

$$v_m = \frac{h\pi\omega}{2\beta} \qquad (3.21)$$

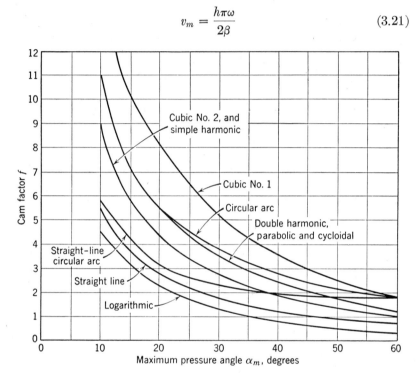

Fig. 3.8. Cam factors f for basic curves. (Also shows cam size comparisons for same maximum pressure angle.)

Substituting eqs. 3.19 and 3.21 in eq. 3.20 gives the cam factor

$$f = \frac{\pi}{2} \cot \alpha_m \qquad (3.22)$$

In Table 3.1 and Fig. 3.8, the equations and values of cam factors for all basic curves are shown.

Cam pitch point location

Cam design generally begins with the location of the pitch point and its pitch-circle radius, R_p. With cylindrical cams (Fig. 3.9), the maxi-

Table 3.1 Cam Factors for Basic Curves

Basic Curve	Cam Factor, f †
Straight line	$\cot \alpha_m$
Straight-line circular arc (circular arc radius = total rise)	$2 \tan \dfrac{\alpha_m}{2} + \cot \alpha_m$
Circular arc	$\cot \dfrac{\alpha_m}{2}$
Simple harmonic motion (SHM)	$\dfrac{\pi}{2} \cot \alpha_m$
Double harmonic*	$2 \cot \alpha_m$
Cycloidal	$2 \cot \alpha_m$
Parabolic	$2 \cot \alpha_m$
Cubic no. 1	$3 \cot \alpha_m$
Cubic no. 2	$\frac{3}{2} \cot \alpha_m$

* Pitch point occurs at cam angle $\beta/2$ for all curves except the double harmonic. For the double harmonic, it is located at cam angle $2\beta/3$.

† α_m = maximum pressure angle, degrees.

mum pressure angle at the inner radius is more than the maximum pressure angle at the outer radius. Usually the difference is within the accuracy of the given data. However, in small cams with large rollers,

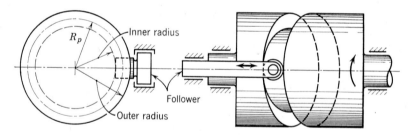

Fig. 3.9. Cylindrical cam. (Note: pitch-circle radius R_p is the mean radius.)

to prevent excessive pressure angles, the value of R_p from eq. 3.19 is suggested for the inner-circle radius. In this manner, the pressure angle will not exceed the limiting value.

Some cam curves have more than one point of maximum slope. The straight-line circular arc displacement diagram (Fig. 3.10) is one of this type. The pitch point is shown. If it were taken at any point higher, the actual maximum pressure angle in the radial cam construction would exceed by distortion the limiting pressure angle α_m.

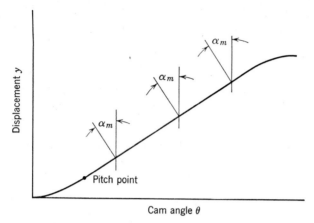

Fig. 3.10. Displacement diagram location of pitch point for combination curves (using straight-line-circular arc as an example).

The proposition for all curves is:

The pitch point shall be the lowest point on the displacement diagram having the greatest slope.

Note that, for the multiple-step cams, the final choice of pitch point is related to the distances between steps.

Example

A cylindrical cam rotates 180 rpm with simple harmonic motion, the follower rising 4 in. in 150 degrees of cam rotation. Find the pitch-circle radius if the pressure angle is to be limited to 35 degrees.

Solution

From Table 3.1, the cam factor

$$f = \frac{\pi}{2} \cot \alpha_m = \frac{\pi}{2} \cot 35 = 2.16$$

From eq. 3.19, the pitch-circle radius

$$R_p = \frac{fh}{\beta} = \frac{2.16(4)}{150 \left(\dfrac{\pi}{180} \right)} = 3.30 \text{ in.}$$

The cylindrical cam may now be constructed on the basis that the pitch point and transition point are both at $\beta/2 = 75$ degrees. The complete solution would be exactly the same for the pitch circle of a radial cam.

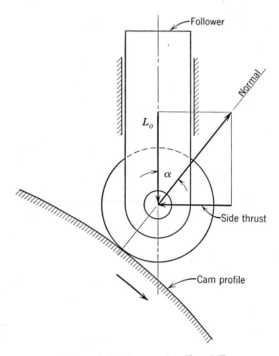

Fig. 3.11. Side thrust of roller follower.

3.10 SIDE THRUST COMPARISON OF CURVES—
RADIAL CAM TRANSLATING ROLLER FOLLOWER

We have shown in Art. 3.4 that the significance of the pressure angle in radial cams with translating roller followers is the side thrust produced on the follower bearing. During action, the side thrust varies continuously for all curves except the logarithmic spiral, which has a constant pressure angle.

Let us compare the curves on the basis of the magnitudes of the side thrust that they produce. All cams have the same conditions: 1-in. lift; 1000 rpm; base-circle radius equals 2.8 in.; total action occurs in 120 degrees of cam rotation; follower linkage weighs 1 lb; and the initial spring force is 20 lb. The tangent cam had an additional dimension: the distance between the cam center and the center of the nose circle equals 2 in. Simplifying the derivations of Art. 3.4 by neglecting guide lengths A and B and the friction, we see from Fig. 3.11

$$\text{Side thrust} = L_o \tan \alpha$$

Figure 3.12[1] presents a comparison of the follower side thrust for six curves: (a) parabolic, (b) simple harmonic motion, (c) cubic no. 1, (d) cycloidal, (e) circular arc-tangent (Fig. 5.11c), and (f) double harmonic. The maximum side thrust of the parabolic curve was

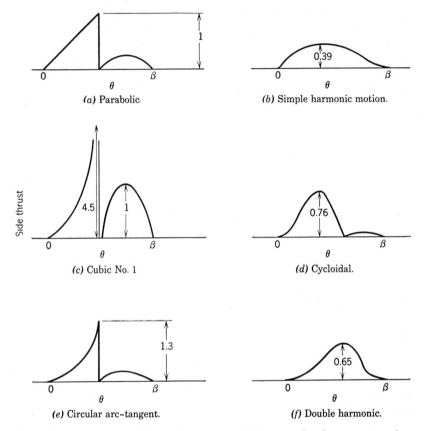

(a) Parabolic

(b) Simple harmonic motion.

(c) Cubic No. 1

(d) Cycloidal.

(e) Circular arc-tangent.

(f) Double harmonic.

Fig. 3.12. Roller cam translating roller follower side thrust comparison of some curves.

assumed as one unit and compared with the other curves. Again, the trigonometric functions, curves b, d, and f, are superior to the others. The simple harmonic motion curve is excellent since inertia forces are high early in the stroke where the pressure angle is small, and low in the middle where the pressure angle is high. Thus, in design a higher maximum pressure angle or smaller cam is allowed with trigonometric curves.

Cam Curvature

3.11 INTRODUCTION

In the previous discussion, we have seen that pressure angle and cam size are directly related, i.e., a limiting pressure angle will determine a certain size cam. However, another condition may exist to preclude the use of a chosen cam size. The curvature of the cam may be too sharp. If this occurs, the follower may not follow the prescribed pitch curve and the stresses on the cam profile may be prohibitive.

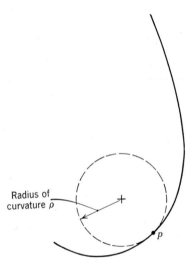

The shape of a curve at any point (its flatness or sharpness) depends on the rate of change of direction called curvature. We may construct for each point of the curve a tangent circle whose curvature is the same as that of the curve at that point. The radius of this circle is called radius of curvature. Figure 3.13 shows the radius of curvature of a point p on the contour. Note that the radius of curvature is continuously changing as we move toward other points on the profile. In this chapter, we shall present relationships for the radial translating follower. Investigation of offset and oscillating followers are mathematically involved, although Shorter[2] shows a tabulation of some mathematical relationships. Therefore, for the offset and oscillating followers, it is suggested that the reader either measure from the layout or use the translating follower data as an approximation. These methods provide reasonable results for properly designed oscillating followers and most offset followers.

Fig. 3.13. Curvature of a contour.

Radius of curvature ρ

3.12 UNDERCUTTING PHENOMENON—ROLLER FOLLOWER

Undercutting is the condition of the constructed or fabricated cam profile that has an inadequate curvature to produce correct follower movement. With a roller follower, undercutting occurs in a convex curve when the radius of curvature of the pitch curve ρ_k is less than the radius of the roller R_r. With a concave curve, it occurs when ρ_k is less than zero or the radius of curvature of the cam profile ρ_c is less than R_r. A sharp point in the pitch curve results.

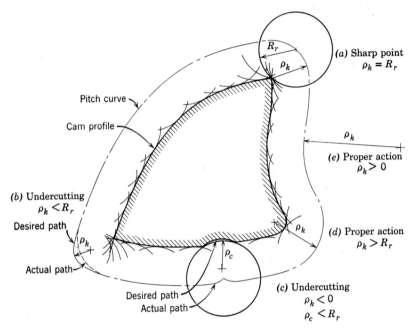

Fig. 3.14. Effect of curvature on cam and follower.

With the convex curve (Fig. 3.14), at point (a), we have the condition $\rho_k = R_r$ so that the center of the roller follower moves on the pitch curve. However, a sharp point in the cam profile is produced. This corner is prohibitive because of high stresses and short life. At the convex curve, point (b), ρ_k of the pitch curve is less than R_r. Undercutting occurs, and the developed cam profile will not constrain the trace point to the pitch curve. The motion of the follower (dotted path) will not be as planned. At the concave curve, point (c), ρ_k is less than zero and a cusp is formed. Again, a form of undercutting occurs in which the follower cannot follow the path desired. But, at points (d) and (e), the radius of curvature ρ_k is adequate in size and the action is proper. The following suggestions are offered to alleviate the condition of undercutting or sharp corners with their inherent high surface stresses:

(*a*) Use a smaller roller diameter. This is limited since the stress in the roller may be excessive.

(*b*) Utilize a larger cam size. This is the easiest solution, but it generally has a practical limitation.

(*c*) Employ an internal cam. The curvature is not as critical in these types since smaller induced surface stresses exist.

As with all design problems, the final choice is a product of intelligent compromise of all the factors involved. For more on this subject of curvature, see references.[5,6,13,14]

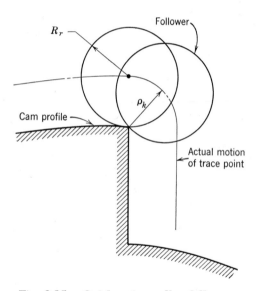

Fig. 3.15.   Quick-action roller follower.

3.13   CONCAVE CAM UNDERCUTTING—TRANSLATING ROLLER FOLLOWER

We shall find that a relationship exists between the follower acceleration and the curvature of a cam. This can be easily shown by consideration of the quick-acting roller follower (Fig. 3.15). We see that the radius of curvature of the pitch curve ρ_k equals the radius of roller R_r. The actual follower motion is a curved path. The acceleration of the follower may be increased by having a smaller path of movement, which means a smaller roller. Increasing the radius of curvature ρ_k decreases the acceleration. Thus, we realize that the radius of curvature of any surface is related to the follower acceleration.

Cam size

In the following we shall indicate a mathematical approach to the minimum radius of curvature of a concave cam profile. This information is needed in establishing the maximum cutter and grinding

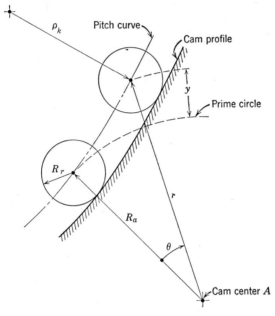

Fig. 3.16. Concave cam profile terminology.

wheel sizes for concave types, such as lobe cams and aircraft cams. In Fig. 3.16, we observe that the radial distance to any point of the pitch curve is

$$r = R_a + y \tag{3.23}$$

where R_a = radius of the prime circle, in. From calculus, the radius of curvature

$$\rho = - \frac{\left[r^2 + \left(\frac{dr}{d\theta} \right)^2 \right]^{3/2}}{r^2 + 2 \left(\frac{dr}{d\theta} \right)^2 - r \frac{d^2r}{d\theta^2}} \tag{3.24}$$

The minus sign indicates the radius of curvature of the concave surface.

Substituting eq. 3.23 in eq. 3.24 and knowing that $dr/d\theta = dy/d\theta$

gives the pitch curve radius of curvature,

$$\rho_k = - \frac{\left[(R_a + y)^2 + \left(\dfrac{dy}{d\theta}\right)^2 \right]^{3\!/\!2}}{(R_a + y)^2 + 2\left(\dfrac{dy}{d\theta}\right)^2 - (R_a + y)\dfrac{d^2y}{d\theta^2}} \quad \text{in.} \quad (3.25)$$

We see from eq. 3.25 that ρ_k is usually a minimum positive value when $dy/d\theta = 0$, which occurs at $y = 0$ and $\theta = 0$. Substituting these conditions yields the minimum pitch curve radius of curvature

$$\rho_{k(\min.)} = - \frac{R_a{}^2}{R_a - \left(\dfrac{d^2y}{d\theta^2}\right)_{\text{at } \theta = 0}} \quad (3.26)$$

Or, in terms of $y = g(t)$, from Art. 2.3

$$\frac{d^2y}{d\theta^2} = \frac{1}{\omega^2}\frac{d^2y}{dt^2} \quad (2.5)'$$

Substituting,

$$\rho_{k(\min.)} = - \frac{R_a{}^2}{R_a - \dfrac{1}{\omega^2}\left(\dfrac{d^2y}{dt^2}\right)_{\text{at } \theta = 0}} \quad \text{in.} \quad (3.27)$$

Note, in Fig. 3.16, that, if eq. 3.27 has positive results for the concave curve, then the minimum concave radius of curvature of the *cam profile* is $\rho_{c(\min.)} = \rho_{k(\min.)} + R_r$. Thus, we may easily calculate the smallest curvature of a concave profile before the layout of the cam.

Grinder size

Frequently in practice it is not the follower-roller diameter that may control the minimum radius of curvature but the size of the larger grinding wheel. This member is always made large compared to the roller diameter for reasonable grinding wear life. Also, the cam curvature must be adequate to permit cutting of the cam with a grinding wheel. From Fig. 3.17, we see that the fundamental basis for cutting the correct profile is that both the roller follower and the grinding wheel lie on the same normal to the cam profile. Also the curvature relationship is

$$\rho_{k(\min.)} + R_r > R_g \quad (3.28)$$

where $R_g =$ radius of the grinding wheel, in. Substituting eq. 3.27

yields

$$-\frac{R_a{}^2}{R_a - \dfrac{1}{\omega^2}\left(\dfrac{d^2 y}{dt^2}\right)_{\text{at }\theta=0}} + R_r > R_g \qquad (3.29)$$

Thus the concave curvature radius of the cam profile shall be greater than the radius of grinding wheel R_g. However, in most cam designs,

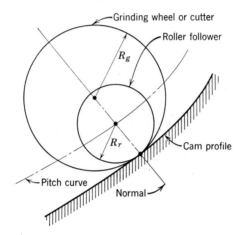

Fig. 3.17. Grinding wheel and roller follower lie on same normal.

the concave contour rarely is smaller than the grinding wheel or roller sizes, and mathematical investigation is not necessary.

Example

A radial cam rotates at 1200 rpm with the translating follower rising ½ in. with simple harmonic motion in 60 degrees of cam rotation. The follower roller is 1¼ in. in diameter, the grinding wheel is 4¾ in. in diameter, and the prime circle is 3¼ in. in diameter. Determine whether the grinding wheel is small enough to cut the cam profile.

Solution

Total cam angle $\qquad \beta = 60\,\dfrac{\pi}{180} = 1.047$ radians

Cam angular velocity $\omega = 1200 \times \dfrac{2\pi}{60} = 125.6$ rad/sec

From eq. 2.20, the acceleration at cam angle $\theta = 0$

$$\frac{d^2y}{dt^2} = \frac{h}{2}\left(\frac{\pi\omega}{\beta}\right)^2$$

$$= \frac{\frac{1}{2}}{2}\left(\frac{\pi \times 125.6}{1.047}\right)^2 = 35{,}500 \text{ in./sec}^2$$

From eq. 3.29, the curvature relationship is

$$-\frac{R_a{}^2}{R_a - \frac{1}{\omega^2}\left(\frac{d^2y}{dt^2}\right)_{\text{at }\theta=0}} + R_r > R_g$$

$$-\frac{(1\frac{5}{8})^2}{1\frac{5}{8} - \left(\frac{1}{125.6}\right)^2 (35{,}500)} + \frac{5}{8} > 2.375$$

The radius of curvature of cam profile equals

$$4.81 \text{ in.} > 2.375 \text{ in.}$$

Therefore, the grinding wheel is small enough or the cam is large enough to fabricate the cam.

3.14   CONVEX CAM UNDERCUTTING—TRANSLATING ROLLER FOLLOWER

Inadequate convex curvature of the cam profile is a frequent problem in cam design. Similar to the last article, the curvature limitation for a convex profile is related to the acceleration of the follower. Again, the control of either curvature or acceleration will determine the limitation of the other, undercutting being the result of too high a negative acceleration. As shown in Art. 3.12, undercutting is undesirable because of incorrect follower action, sharp cam profile, and high surface stresses. In Fig. 3.18, we see a cam profile and pitch curve in which the roller follower has a negative acceleration. With arc DE a finite value, the prevention of undercutting requires that the minimum radius of curvature of the pitch curve

$$\rho_{k(\text{min.})} > R_r \tag{3.30}$$

Therefore, using eq. 3.25 with a positive sign since the curvature is toward the center of the cam, we have

$$\frac{\left[(R_a + y)^2 + \left(\frac{dy}{d\theta}\right)^2\right]^{3/2}}{(R_a + y)^2 + 2\left(\frac{dy}{d\theta}\right)^2 - (R_a + y)\frac{d^2y}{d\theta^2}} > R_r \tag{3.31}$$

Converting to functions of time gives

$$\frac{\left[(R_a + y)^2 + \left(\frac{dy}{dt} \times \frac{1}{\omega}\right)^2\right]^{3/2}}{(R_a + y)^2 + 2\left(\frac{dy}{dt} \times \frac{1}{\omega}\right)^2 - (R_a + y)\left(\frac{d^2y}{dt^2} \times \frac{1}{\omega^2}\right)} > R_r \quad (3.32)$$

Generally, the smallest radius of curvature occurs at the point of maximum negative acceleration. Therefore, to solve eq. 3.31 or eq. 3.32, use the values of displacement, velocity, and acceleration at the point of maximum negative acceleration.

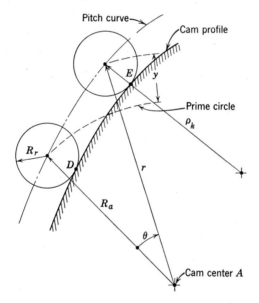

Fig. 3.18. Convex cam profile terminology.

Often the radius of curvature is needed to determine the contact stresses between cam and follower. The larger the radius of curvature, the smaller are the surface stresses. Thus, good practice requires that $\rho_{k(\min.)}$ exceed R_r by a reasonable margin of safety. If $\rho_k = R_r$, we would have an impractical, sharp point on the cam profile.

Example

A radial cam rotates at 1200 rpm with the follower rising 1 in. with simple harmonic motion in 150 degrees of cam rotation. The roller is $1\frac{1}{4}$ in. in diameter, and the prime circle is $3\frac{1}{4}$ in. in diameter. Calculate whether there will be undercutting.

Solution

Total cam angle $\beta = 150 \times \dfrac{\pi}{180} = 2.62$ radians

Cam angular velocity $\omega = 1200 \times \dfrac{2\pi}{60} = 125.6$ rad/sec

The point of maximum negative acceleration for simple harmonic motion occurs from eq. 2.20 at cam angle β, giving

$$\left(\frac{d^2y}{dt^2}\right)_{\text{max. neg.}} = -\frac{h}{2}\left(\frac{\pi\omega}{\beta}\right)^2$$

$$= -\frac{1}{2}\left(\frac{\pi \times 125.6}{2.62}\right)^2 = -11{,}400 \text{ in./sec}^2$$

The displacement at this angle β is $y = 1$ in. The velocity dy/dt at this angle β, from eq. 2.19, is equal to zero.

From eq. 3.32, the radius of curvature is

$$\frac{\left[(R_a + y)^2 + \left(\dfrac{dy}{dt} \times \dfrac{1}{\omega}\right)^2\right]^{3/2}}{(R_a + y)^2 + 2\left(\dfrac{dy}{dt} \times \dfrac{1}{\omega}\right)^2 - (R_a + y)\left(\dfrac{d^2y}{dt^2} \times \dfrac{1}{\omega^2}\right)} > R_r$$

Substituting,

$$\frac{[(1\frac{5}{8} + 1)^2 + 0]^{3/2}}{(1\frac{5}{8} + 1)^2 + 0 - (1\frac{5}{8} + 1)\left(\dfrac{-11{,}400}{(125.6)^2}\right)} > \frac{5}{8}$$

Thus, the radius of curvature of the pitch curve equals

$$2.16 \text{ in.} > \frac{5}{8} \text{ in.}$$

Therefore, there is no undercutting and the cam profile is correct, giving proper follower movement. It may be mentioned that the minimum radius of curvature of the cam profile equals $2.16 - 0.625 = 1.54$ in. For a more complete analysis, the pressure angle should be investigated for this cam size.

3.15   UNDERCUTTING PHENOMENON—FLAT-FACED FOLLOWER

Since the pressure angle has only a minor effect on the flat-faced follower (Art. 3.5), a factor limiting the cam size of this follower is undercutting. In Fig. 3.19a a flat-faced follower is located at its respective positions 1, 2, and 3 to construct the cam. The cam profile, drawn tangent to the flat-faced follower, cannot be made to contact the follower at position 2. In other words, follower positions

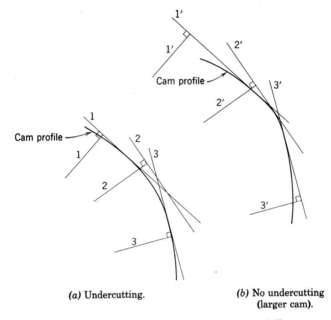

(a) Undercutting.   (b) No undercutting (larger cam).

Fig. 3.19. Undercutting with a flat-faced follower.

1 and 3 eliminate or undercut position 2. The cam thus developed is incapable of driving the follower in the desired manner. Figure 3.19b shows a larger cam without undercutting. Thus, undercutting limits the cam size to a minimum value.

3.16 UNDERCUTTING—TRANSLATING FLAT-FACED FOLLOWER

A new approach must be found to establish mathematically the flat-faced follower cam curvature, since substituting the value of infinity for R_r in eq. 3.31 gives no results.

In Fig. 3.20, we see this follower in contact with the cam. Let

R_b = radius of base circle, in.

q = eccentricity of contact point from cam center, in.

u = distance from centro to cam center of curvature, in.

From kinematics, the centro or center of rotation between two bodies is a point common to these bodies and has the same velocity relative to the ground considered on either body. Therefore, the velocity of the

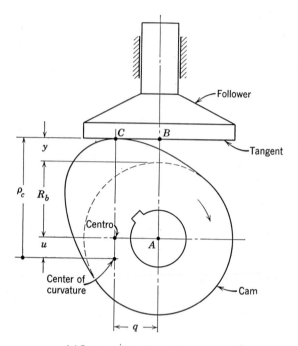

(a) Large radius of curvature ρ_c.

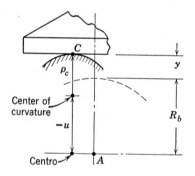

(b) Small radius of curvature ρ_c.

Fig. 3.20. Cam curvature limitation with translating flat-faced follower.

translating follower or centro is

$$v = \frac{dy}{dt} = q\omega \tag{3.33}$$

The center of curvature of the cam at this instant is assumed a distance

u from the centro. And the velocity of this point in direction of the tangent is

$$\frac{dq}{dt} = u\omega \tag{3.34}$$

The acceleration of the follower, differentiating eq. 3.33, is

$$a = \frac{d^2y}{dt^2} = \frac{dq}{dt}\,\omega \tag{3.35}$$

Substituting eq. 3.34 in eq. 3.35 gives the vertical distance

$$u = \frac{1}{\omega^2} \times \frac{d^2y}{dt^2} \tag{3.36}$$

We now see that the location of the center of curvature is directly related to the follower acceleration. Practically speaking, for flat-faced follower cam construction there is no limitation of curvature (difficulty in layout) during the positive acceleration period. This requires that the distance u is positive below the centro (Fig. 3.20a). The restriction occurs when the acceleration is negative, which gives u a minus value (Fig. 3.20b). To eliminate sharp corners and undercutting in the cam profile, the minimum radius of curvature shall equal

$$\rho_{c\,(\text{min.})} = R_b + \underset{\text{max. neg.}}{(y + u)} > 0 \tag{3.37}$$

Or

$$\underline{\rho_{c(\text{min.})} = R_b + \left(y + \frac{1}{\omega^2}\frac{d^2y}{dt^2}\right)_{\text{max. neg.}} > 0} \tag{3.38}$$

Since R_b and y are always positive, the undercutting ($\rho_{c(\text{min.})} < 0$) is due to excessive negative acceleration. Also, as mentioned previously, a margin of safety is necessary in the equations to keep the curvature reasonable and the stresses low. If small curvature or undercutting exists, then we see from eq. 3.38 that it is necessary to employ either a larger cam or a smaller maximum negative acceleration. Note that eq. 3.38 may be used to ascertain the cam curvature of any point by substituting the respective parameter values. It may be mentioned that the cam profile cannot always be drawn for all functions of y.

Example

A radial cam rotates at 1200 rpm with the translating flat-faced follower rising 1 in. with simple harmonic motion in 150 degrees of cam rotation. The base-circle diameter is 3 in. Calculate whether the cam is large enough to preclude undercutting.

Solution

Cam angular velocity $\omega = 1200 \times \dfrac{2\pi}{60} = 125.6$ rad/sec

Total cam angle $\quad \beta = 150 \dfrac{\pi}{180} = 2.62$ radians

The point of maximum negative acceleration for simple harmonic motion from eq. 2.20 occurs at cam angle β, giving

$$\left(\frac{d^2y}{dt^2}\right)_{\text{max. neg.}} = -\frac{h}{2}\left(\frac{\pi\omega}{\beta}\right)^2$$

$$= -\frac{1}{2}\left(\frac{\pi \times 125.6}{2.62}\right)^2 = -11{,}400 \text{ in./sec}^2$$

The displacement at this angle β is $y = 1$ in. Therefore, from eq. 3.38, the minimum radius of curvature of the cam profile

$$\rho_{c(\text{min.})} = R_b + \left(y + \frac{1}{\omega^2}\frac{d^2y}{dt^2}\right)_{\text{max. neg.}} > 0$$

$$= 1\tfrac{1}{2} + 1 + \frac{-11{,}400}{(125.6)^2} > 0$$

$$= 1.78 \text{ in.} > 0$$

Therefore, no undercutting is evident and the cam will be a proper design if the stresses are not excessive.

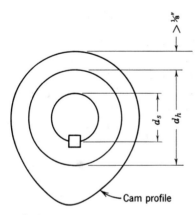

Fig. 3.21. Hub size.

3.17 HUB DESIGN

Obviously, the cam must have a hub large enough to accommodate the shaft upon which it turns. The first step in the design of a cam-follower system is to establish the size of cam shaft and the hub key necessary. Stresses and deflections are the controlling factors in this choice. For the cast-iron cam, the hub diameter is

$$d_h = 1\tfrac{3}{4}d_s + \tfrac{1}{4} \text{ in.} \quad (3.39)$$

where d_s is the shaft diameter in inches (Fig. 3.21).

For steel cams, the hub may be made slightly smaller. For additional strength, locate the key in the largest portion of the cam body, which will also easily establish the proper relative position of the follower and the cam in assembling the mechanism. In practice, the cam profile is made larger than the hub by at least $\frac{1}{8}$ in. This is to prevent stress concentrations and short life in the cam surface.

3.18 SUMMARY

The three pertinent factors involved in determining the minimum cam size are pressure angle, profile curvature, and hub size. The pressure angle is the steepness of the cam and as such determines the side thrust or wear of the translating roller follower in its bearing guide and the performance of an oscillating roller follower. This side thrust only exists during the rise, i.e., when the follower motion is in the opposite direction to the follower load. The maximum pressure angle should be made as small as possible during this action. For the roller follower, this is done by having the coefficient of friction and the follower stem overhang small, and the follower bearing long with the stem as rigid as possible. With flat-faced followers, the pressure angle, being constant (usually equal to zero), is not of design consideration. Also, the side-thrust pressure angle effects do not exist with properly designed oscillating followers. For most machines having translating roller followers, the maximum pressure angle is 30 degrees or less. But, the author has gone as high as $47\frac{1}{2}$ degrees with light loads and rigid, low-friction linkages. Reduction in pressure angle may be achieved by using a larger cam, an offset follower, a different cam curve, or a secondary follower. The cam size is limited by the maximum pressure angle, the chosen curve, the total rise, and the cam angle for this rise. Furthermore, transferring the pitch curve from the displacement diagram to a radial cam produces distortion, which changes the pressure angle. Equation 3.19 may be used to establish the roller follower cam size for cylindrical cams and for yielding acceptable accuracy for radial cams.

We have seen that all cams cannot be constructed to give the desired motion to the followers. This condition, called undercutting, is attributed to the follower surface radius of curvature, the follower acceleration and the cam size. In concave profiles using roller followers, the acceleration at $\theta = 0$ generally controls. In convex profiles, using either a roller or flat-faced follower, the maximum negative acceleration controls. The easiest solution to this problem of undercutting is to employ a larger cam. Furthermore, with or without undercutting, sharp corners may be produced in the cam profile. These are undesirable since high stresses and low cam life result.

Lastly, the hub or shaft size is another limiting factor in determining the minimum cam size.

References

1. J. Jennings, "High Speed Cam Profiles," *Mach., London 52*, p. 521 (July 28, 1938).
2. W. H. Shorter, "Cam Design Analysis," *Prod. Eng. 11*, p. 223 (May 1940).
3. J. A. Hrones, "Analysis of Dynamic Forces in a Cam Driven System," *Trans. ASME 70*, p. 473 (1948).
4. M. Kloomok and R. V. Muffley, "Determination of Pressure Angles for Swinging-Follower Cam Systems," *ASME Paper No. 55–SA–38*, 1955.
5. W. M. Dudley, "New Methods in Valve Cam Design," *Trans. SAE 2*, p. 19 (Jan. 1948).
6. M. L. Baxter, "Curvature Acceleration Relation for Plane Cams," *Trans. ASME 70*, p. 483 (1948).
7. A. H. Candee, "Kinematics of Disk Cam and Flat Follower," *Trans. ASME 69*, p. 709 (1947).
8. W. B. Carver and B. E. Quinn, "An Analytical Method of Cam Design," *Mech. Eng. 67*, p. 523 (Aug. 1945).
9. A. R. Holowenko and A. S. Hall, "Cam Curvature," *Mach. Des. 25*, p. 170 (Aug. 1953), p. 162 (Sept. 1953), p. 148 (Nov. 1953).
10. W. Richards, "Cam Design—General Principles and Classification of Cams," *Mach., London 48*, p. 169 (May 7, 1936).
11. M. Kloomok and R. V. Muffley, "Determination of Radius of Curvature for Radial and Swinging-Follower Cam Systems," *ASME Paper No. 55–SA–89*, 1955.
12. W. Hartman, "Ein Neues Verfahren zur Aufsuchung des Kruemmungskreises," *Zeitschrift VDI 37*, p. 95 (1893).
13. N. Rosenauer and A. H. Willis, *Kinematics of Machines,* Associated General Publications Ltd., Sydney, Australia, p. 38, 1953.
14. R. Beyer, *Kinematische Getriebesythese,* Springer, Berlin, Germany, p. 46, 1953.

Cam Profile Determination

4.1 INTRODUCTION

In this chapter we shall establish methods for determining the cam shape with the profile plotted as accurately as possible. Since this layout procedure has limited precision when used as a means for fabricating the contour, we shall also show calculations in which the profile and machine tool cutter location are determined exactly. This results in the high fabrication accuracy that is preferred for moderate- or higher-speed cams. For a discussion of accuracy, the reader is referred to Chapter 10.

In cam construction, we shall utilize the information of the previous chapters. The pressure angle, curvature, undercutting, and practical factors of follower location and sizes are determined until the ultimate cam contour is established.

As previously stated the fundamental basis for all cam layouts is that

the cam profile is developed by fixing the cam and moving the follower around the cam to its respective relative positions.

In this manner, we maintain the same cam-follower relative motion as the cam mechanism. If the reader remembers this tenet, the layout procedures will be easily understood and retained. For clarity, only a few layout points will be shown. However, since accuracy of cam shape is essential many points are needed in constructing the actual cam. Enough points must be taken to establish the contour with confidence. In Fig. 4.1, we see a roller follower plotted for an infinite number of positions of which the cam profile will be the envelope shown. The same procedure can be followed analytically (by finding the parametric equations of the envelope) but the mathematics is generally tedious.

The first step in cam construction is the establishment of the reference point (cam size) which relates the displacement diagram

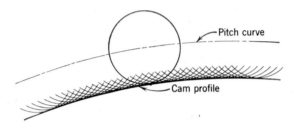

Fig. 4.1.   Envelope of roller follower produces a cam profile.

to the cam layout. It is either the trace point or a point on the flat-faced follower surface. With reference to a roller or spherical follower, we see from Fig. 4.2 that the cam profile ab is drawn tangent

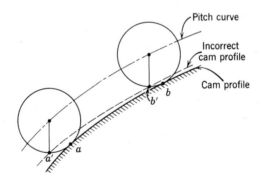

Fig. 4.2.   Proper roller-follower cam profile.

to the roller after the pitch curve is determined. It would be wrong to employ as the reference point the bottom point of the roller since this would give the incorrect profile $a'b'$ and thus a different cam.

4.2   RADIAL CAM—TRANSLATING KNIFE-EDGE FOLLOWER

Let us begin with the simplest case: a knife-edge follower. This type of follower, although impractical, is presented to indicate the fundamentals of cam construction.

In Fig. 4.3b, we are given a pitch curve of the displacement diagram having the following motion: (a) rise of $\frac{5}{8}$ in. in 150 degrees of cam rotation; (b) dwell for 60 degrees; (c) fall of ⅝ in. in 120 degrees; (d) dwell for the remaining 30 degrees.

A translating follower is employed with the cam rotating clockwise having a base-circle radius R_b of ¾ in. The procedure for layout is as follows:

(a) Starting with the reference point 0' on the line of follower motion and the base circle, locate the point 0 of the displacement diagram opposite 0'.

(b) On the displacement diagram, divide the rise angle of 150 degrees into equal parts, i.e., 6 parts of 25 degrees each, giving points 1, 2, 3, etc.

(c) In Fig. 4.3a, divide the cam angle of 150 degrees into the same number of radial lines $A0''$, $A1''$, $A2''$, etc., i.e., 6 lines 25 degrees apart.

(d) Project points 1, 2, 3, etc., from the displacement diagram to the cam line of the follower motion, giving rise points 1', 2', 3', etc.

(e) With cam center at A, swing arcs 1–1', 2–2', 3–3', etc., intersecting the respective radial lines, giving points on pitch surface 0, 1, 2, 3, etc.

(f) Draw a smooth curve through these points, which gives this portion of the cam profile.

We can continue in the same manner to locate other points and complete the cam profile. The dwell portions of the cam are at a constant radial distance. Note that a pair of dividers could have been used to give the rise at each point in lieu of projecting lines from the displacement diagram to the cam. In the foregoing example, no effort was made to control the cam pressure angle, follower velocity, or acceleration. These items are determinable.

4.3 RADIAL CAM—TRANSLATING ROLLER FOLLOWER

In this article, we shall be more practical and use a roller follower. Also we shall indicate the direct approach for a cam layout with the pressure angle held within practical limits. The reference point will be the trace point on the line of follower motion (Fig. 4.4). The reader should refer to the previous chapter for limitations on the roller and cam sizes.

Example

A radial cam rotating at 180 rpm is driving a ¾-in.-diameter translating roller follower. The motion is as follows: (a) rise of 1 in. with simple harmonic motion in 150 degrees of cam rotation; (b) dwell for 60 degrees; (c) fall 1 in. with simple harmonic motion in 120 degrees of cam rotation; (d) dwell for the remaining 30 degrees.

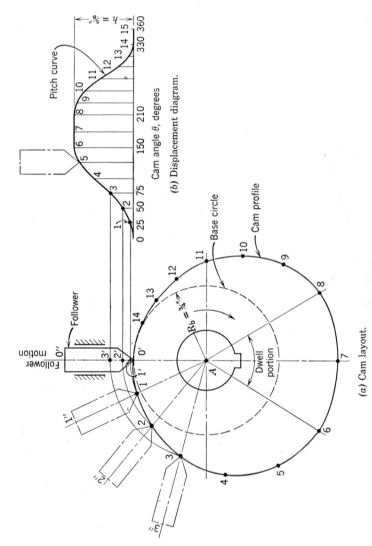

(a) Cam layout.

(b) Displacement diagram.

Fig. 4.3.   Radial cam-translating knife-edge follower—full scale.

Construct the cam contour two ways: (1) with the pressure angle limited to approximately 20 degrees, and (2) with the pressure angle limited to exactly 20 degrees.

Solution—Approximate Pressure Angle Method

(a) Make a freehand sketch of the displacement diagram (Fig. 4.4b), showing the cam angles and rise.

(b) Find the reference point. In Art. 3.9, it was assumed that the pitch point was located at the transition point cam angle $\beta/2$ or 75 degrees. Equation 3.19 gives the pitch-circle radius, $R_p = 1.64$ in. Let us take $1\frac{3}{4}$ in. as a safe dimension. Thus, in Fig. 4.4a, we have the location of the pitch point from the cam center A.

(c) Divide the cam angle of 150 degrees into equal parts, i.e., 6 parts at 25 degrees on either side of the line of action $A3''$. This gives radial lines $A0''$, $A1''$, $A2''$, etc.

(d) Construct the simple harmonic motion on line $A3''$ with the center of a 1-in. circle at the pitch point. This circle is divided into the same number of equal parts, each giving, when projected, points $0'$, $1'$, $2'$, $3'$, $4'$, etc.

(e) From the cam center A, swing arcs from points $0'$, $1'$, $2'$, $3'$, etc., intersecting respective radial lines, locating trace points 0, 1, 2, 3, etc.

(f) Draw the smooth pitch curve, roller circles, and cam profile tangent to the rollers.

(g) Continue in the same manner to complete the cam.

Note that the *actual* maximum pressure angle can be measured on the cam. It occurs on the pitch curve between points 2 and 3.

Solution—Exact Pressure Angle Method

In constructing a cam with an exact pressure angle limitation, utilize the methods of Art. 3.8 to give the radius of pitch circle, $R_p = 1.565$ in., occurring at a cam angle $\theta_p = 60.9$ degrees. In 60.9 degrees, the rise of the follower is given by eq. 2.18

$$y = \frac{h}{2}\left(1 - \cos\frac{\pi\theta}{\beta}\right)$$

$$= \frac{1}{2}\left(1 - \cos\frac{\pi \times 60.9}{150}\right) = 0.354 \text{ in.}$$

(a) In Fig. 4.4c, draw the pitch-circle radius $R_p = 1.565$ in., locating the pitch point. Since $A0''$ is 60.9 degrees from the pitch point, we can draw the radial lines $A0''$, $A1''$, $A2''$, etc.

(b) The lowest point of the rise is 0.354 in. below the pitch point, which locates points $0'$ and $3'$ on its 75-degree radial line $A3''$.

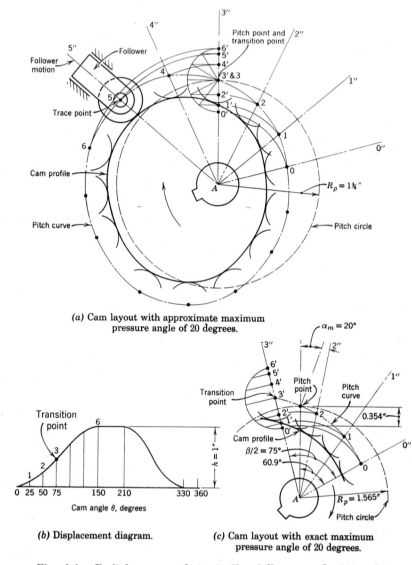

(a) Cam layout with approximate maximum pressure angle of 20 degrees.

(b) Displacement diagram.

(c) Cam layout with exact maximum pressure angle of 20 degrees.

Fig. 4.4. Radial cam-translating roller follower, scale $\frac{5}{8}'' = 1''$.

(c) The procedure that follows is similar to the previous approximate pressure angle method.

By measurement, we see that the maximum pressure angle occurs at the pitch point and is *exactly* 20 degrees. Note that the actual cam size in either event may have to be increased by the hub or shaft size, stresses in roller or cam due to small radius of curvature, and other interrelated factors. However, concerning adequate cam curvature, the layout shows that no undercutting prevails.

4.4 RADIAL CAM—TRANSLATING FLAT-FACED FOLLOWER

Let us now develop the cam profile for a translating flat-faced follower. We know that this cam-follower mechanism has a constant pressure angle which measures zero degrees for the 90-degree follower face. Therefore, the cam may be as small as possible without serious pressure angle side thrust effect. However, the undercutting condition should always be investigated, especially with small cams.

90-Degree face

Let us construct a radial cam with a 90-degree flat-faced follower having a base-circle diameter of 2 in. The motion is as follows: (*a*) rise of ¾ in. with parabolic motion in 150 degrees of cam rotation; (*b*) dwell for 60 degrees; (*c*) fall of ¾ in. with parabolic motion in 150 degrees of cam rotation completing the action.

Before we start the layout, calculations according to Art. 3.16 should be made. These indicate that the base circle is large enough and undercutting does not exist.

The construction (Fig. 4.5) is as follows:

(*a*) Sketch the displacement diagram.

(*b*) Divide the rise angle of 150 degrees into equal parts to give radial lines $A0''$, $A1''$, $A2''$, etc.

(*c*) On the cam, locate the base-circle reference point $0'$ on the line of action $A0''$.

(*d*) Similar to Art. 2.12, construct the parabolic motion by drawing any line $0B$ which is then divided into the same number of equal parts, i.e., 6 parts spaced 1, 3, 5, 5, 3, 1.

(*e*) These lines are projected to radial line $A0''$, giving rise points $1'$, $2'$, $3'$, etc.

(*f*) From the cam center A, swing arcs locating points 1, 2, 3, etc., of the flat-faced follower movement on the respective radial lines.

(*g*) Through these points, draw the 90-degree flat-faced follower.

(*h*) Construct the cam profile tangent to these flat faces.

By this method, the displacement diagram serves only as a guide with no direct values to be plotted. We could have drawn the diagram accurately and taken cam displacement ordinate values from it.

From the layout, let us determine the smallest size of the mush-room follower face to maintain contact with the cam profile at all times. In other words, the cam should not go over the edge of the

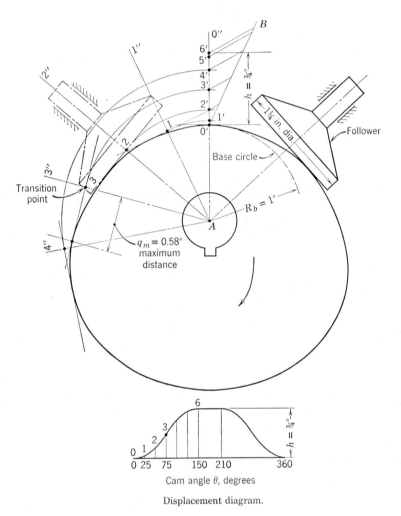

Fig. 4.5.   Radial cam—translating 90-degree flat-faced follower.   Full scale.

follower. We see that position 3 is the critical location (measuring 0.58 in.) where the contact point is farthest from the line of motion. For practical reasons, we shall make the follower radius ⅝ in., giving a mushroom diameter of 1¼ in. It can be shown that the transition point is the critical location in establishing the necessary follower size. Solution for this size may be mathematically established from eq. 3.33. The maximum eccentricity of the point of contact from the cam center is

$$q_m = \frac{v_m}{\omega} \tag{4.1}$$

where v_m = maximum velocity of follower, ips.

ω = cam angular velocity, radians/sec.

This is a convenient relationship which states that the follower face may be made smaller by reducing the maximum follower velocity and increasing the cam speed.

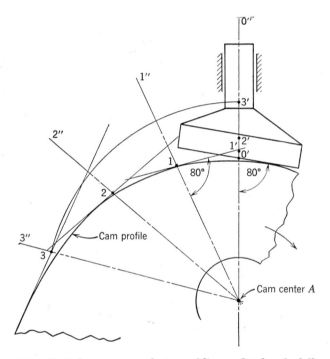

Fig. 4.6. Radial cam—translating oblique flat-faced follower. Scale 1⅝″ = 1″.

Oblique face

Now, let us find the cam profile for a flat-faced follower having a face at an angle other than 90 degrees with the follower motion. In this example, we shall assume the same data as previously, but with a follower face angle of 80 degrees (Fig. 4.6).

The solution is similar to the previous one except that step (*g*) will read

(*g*) Through these points, draw the 80-degree flat-faced follower.

The example of the oblique-faced follower is presented primarily to illustrate fundamentals. The author has never seen any practical application or need for this type of follower. The method of layout discussed in this article may also be used with the oscillating flat-faced follower in which the included angle would change for every relative position of the cam and follower. Note that again the procedure would be similar if the follower face were curved in any way.

4.5 RADIAL CAM—TRANSLATING SPHERICAL-FACED FOLLOWER

The construction procedure of the cam having a spherical-faced follower of relatively small curvature (Fig. 4.7) is similar to the roller follower method of Art. 4.3. In both instances, the pitch curve

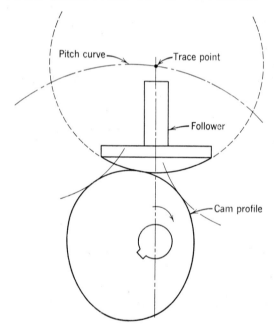

Fig. 4.7.   Radial cam—translating spherical-faced follower.

is the center of curvature of the follower (trace point) and the cam profile is determined by drawing the follower face at its respective positions. It should be noted in Fig. 4.7 that the spherical radius is shown smaller than that generally utilized in practice. Usually the radius varies from 30 in. to 300 in., which gives a relatively flat follower surface. This necessitates a construction method similar to that indicated in Art. 4.4.

4.6 RADIAL CAM—TRANSLATING, OFFSET ROLLER FOLLOWER

We have seen that properly offsetting the follower has the advantage of decreasing the pressure angle side thrust effect. However, contrary to the radial follower, the follower path is not the profile displacement of the cam. In the following example, a roller follower will be employed, the method being similar for all translating offset followers.

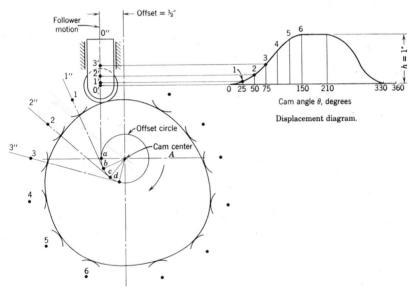

Fig. 4.8. Radial cam—translating, offset roller follower. Scale 6 in. = 1 ft.

In Fig. 4.8, we are given a cam having a ¾-in.-diameter roller follower offset ½ in. and shown in the lowest position 0′. The displacement diagram is also given. Construct the cam profile. The procedure is as follows:

(a) Similar to the previous articles, divide the total cam angle of 150 degrees into equal parts, i.e., 6 parts of 25 degrees each, giving points 1, 2, 3, etc., on the displacement diagram.

(b) Locate point 0 of displacement diagram opposite point 0′, and project points 1′, 2′, 3′, etc., on the line of follower motion 0″a.

(c) Draw an offset circle tangent to the line of motion a0″.

(*d*) Draw lines *b1″*, *c2″*, *d3″*, etc., tangent to offset circle in the same number of parts, i.e., 6 parts at 25-degree increments. This gives the location of the line of motion at its respective positions.

(*e*) With distance *a1′* measured from *b* on line *b1″*, locate the trace point 1.

(*f*) Similarly, with distance *a2′* measured from point *c*, find the trace point 2 on line *c2″*.

(*g*) Continue for all points—draw rollers with the cam profile tangent to them.

There is another method for finding cam profile by using radial lines from the cam center in lieu of the offset circle. It will be shown in another example.

4.7   RADIAL CAM—OSCILLATING ROLLER FOLLOWER

The primary advantage of the oscillating follower is its improved force distribution compared to its translating counterpart. In other words, the pressure angle may be considerably increased in size without detrimental follower action. The follower pivot location with respect to the forces and the direction of cam rotation is an important consideration. Poor choice may produce unsatisfactory action and sometimes locking of the oscillating follower.

The cam layout procedure differs from that shown with the previous translating follower. Since the follower oscillates through an established angle, its required movement is either approximated on an arc of movement or exactly related to the angles of movement. Therefore, as the ordinate in the displacement diagram, we use the arc of movement or angle of movement in lieu of a rectilinear rise.

In the following example (Fig. 4.9), we are given the roller size, prime-circle diameter, total angle of oscillation ϕ_0, occurring in 180 degrees of cam rotation, and the pivot point location with respect to the cam center. The construction procedure (approximated on the arc of movement) is as follows:

(*a*) With the total angle of oscillation ϕ_0 from Fig. 4.9, plot arc length 0′, 3′, 6′, as an ordinate in the displacement diagram. Use small chords of the arc.

(*b*) Select a basic curve, and draw the pitch curve on the displacement diagram.

(*c*) Divide the displacement diagram rise into equal cam angle divisions, e.g., 6 parts at 30 degrees apart, giving points 1, 2, 3, etc.

(*d*) Plot the respective rise divisions 1′, 2′, 3′, etc., on the cam arc of movement 0′, 3′, 6′. Use small chordal lengths.

(*e*) Divide the cam angle of 180 degrees into the same number of equal angles, e.g., 6 parts giving radial lines $A1''$, $A2''$, $A3''$, etc., 30 degrees apart.

(*f*) From the cam center A, draw arcs with radii $A1'$, $A2'$, $A3'$, etc., giving intercepts at a, b, c, d, and a', b', c', d'.

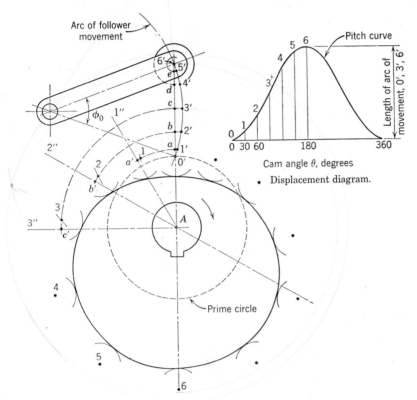

Fig. 4.9.   Radial cam—oscillating roller follower.

(*g*) With chord aa' at $1'$ on the arc of action, swing an arc giving trace point 1. Repeat with chord bb' at point $2'$; swing arc giving point 2, etc.

(*h*) Draw the pitch curve not shown, the rollers, and the cam profile tangent to the rollers. The return action is similar.

We see that the follower is shifted in the direction of cam rotation distances, $a1'$, $b2'$, $c3'$, etc. In other words, the trace points are not on radial lines $A1''$, $A2''$, $A3''$, etc.

4.8   RADIAL CAM—OSCILLATING FLAT-FACED FOLLOWER

The oscillating flat-faced follower differs from its translating counterpart in that its face is changing its angular position with respect to the cam at all times.

In the following example (Fig. 4.10), we are given the base-circle diameter of 4 in., the total angle of oscillation $\phi_0 = 15$ degrees in 180 degrees of cam rotation, and the pivot point location with respect to the cam center. The follower is to be oscillated with parabolic motion.

The layout procedure is:

(a) Divide the angle of oscillation ϕ_0 into equal parts, e.g., 6 parts of parabolic motion increments according to Art. 2.12. This is shown in Table 4.1.

Table 4.1

Points	Increments	Angular Increments, degrees
0		
	1	$\frac{1}{18} \times 15 = \frac{5}{6}$
1		
	3	$\frac{3}{18} \times 15 = \frac{5}{2}$
2		
	5	$\frac{5}{18} \times 15 = 4\frac{1}{6}$
3		
	5	$\frac{5}{18} \times 15 = 4\frac{1}{6}$
4		
	3	$\frac{3}{18} \times 15 = \frac{5}{2}$
5		
	1	$\frac{1}{18} \times 15 = \frac{5}{6}$
6		
	Sum = 18	Sum = 15 degrees

(b) From the follower pivot point B, draw any arc of motion at a radius R_1, and also tangent radius r_1. Locate the increment angles, giving points 0, 1, 2, 3, 4, etc.

(c) Divide the 180-degree cam angle into the same number of equal parts, e.g., 6 parts located 30 degrees apart. These are measured from radial line $A0''$, giving radial lines $A1''$, $A2''$, $A3''$, etc.

(d) With radius AB, draw a circle from the cam center A intersecting these radial lines to give the respective positions of the follower pivot.

(e) From the cam center A, draw arcs 11', 22', 33', etc.

(f) On radial line $A1''$, draw radius R_1, intersecting arc 11' at point 1'.

(g) Draw radius r_1 at the pivot location and then a straight-line tangent to the r_1 circle to point 1'. Repeat tangents for all other points. This gives the respective follower surface positions.

(h) Construct the cam profile tangent to the surfaces.

This method clearly shows the follower being placed around the fixed cam at its respective positions. Although the parabolic curve was chosen, the same procedure utilizing oscillating increment angles may be employed for any other curve, such as the simple harmonic motion, cycloidal, and cubic.

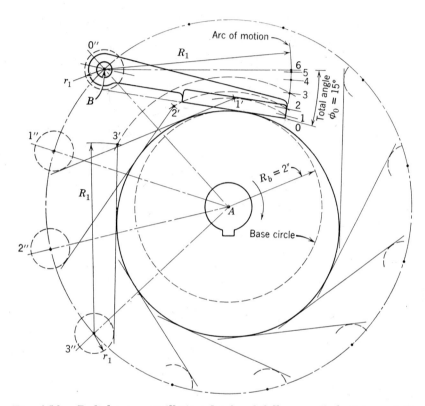

Fig. 4.10.   Radial cam—oscillating flat-faced follower.   Scale 6 in. = 1 ft.

4.9   RADIAL CAM—PRIMARY AND SECONDARY FOLLOWER

We have shown that an excessive pressure angle produces jamming of the translating follower in its guide. This condition may be alleviated by using a primary and a secondary follower (Fig. 4.11). This mechanism allows a smaller cam and a larger pressure angle. The procedure for layout is similar to that shown in the previous articles. The only difference is that the desired movement is shown at the end of a secondary follower and must be referred to the primary follower.

In the following example (Fig. 4.11), the secondary translating follower is to rise $1\frac{1}{4}$ in. with simple harmonic motion in 180 degrees of clockwise cam rotation. The mechanism is shown to scale in its lowest position in contact with a 3½-in.-diameter prime circle. The layout procedure is:

(a) On the secondary follower, divide a 1-in. simple harmonic semicircle into equal parts, e.g., 6 parts giving rise points 0, 1, 2, 3, etc., at the trace point of the follower face (radius ρ_f). Construct dotted arcs 0, 1, 2, 3, etc.

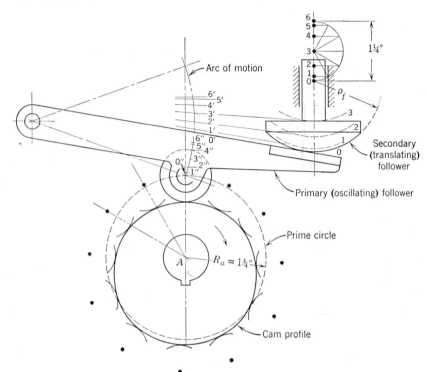

Fig. 4.11.   Radial cam—primary and secondary follower.   Scale 6 in. = 1 ft.

(b) Keeping the primary follower tangent to these arcs gives points 0', 1', 2', 3', etc., on its roller arc of motion.

(c) By proportion, relate these points to movements of the trace point 0'', 1'', 2'', etc., of the primary follower.

(d) Continue by repeating steps (e) through (h) of Art. 4.7, since at this stage the cams are similar.

4.10   FACE-GROOVE RADIAL CAM (POSITIVE DRIVE)

In all the previous cams, it was assumed that the follower was held in contact with the cam by its own weight or by a preloaded compression spring. Now, two contact surfaces guiding a roller will provide follower constraint (Fig. 4.12). As usual, roller positions are drawn after the pitch curve is plotted. However, in this cam type, we construct two tangents to the roller circles, giving two cam profiles.

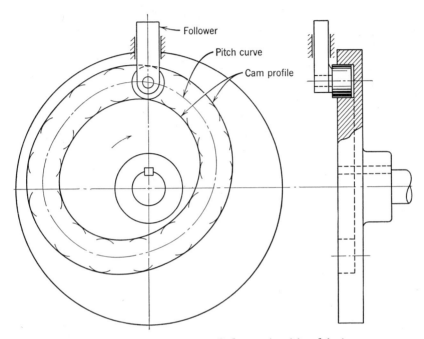

Fig. 4.12. Face-groove radial cam (positive drive).

Later articles will show that the necessary clearance between the roller and the cam profiles may produce unsatisfactory cam-follower action, especially at high speeds. One of the advantages of the grooved disk radial type is that it is one of the best balanced cams. High-speed performance is obviously improved in this respect.

4.11 YOKE RADIAL CAM—FLAT-FACED FOLLOWER (POSITIVE DRIVE)

The yoke radial cam (Fig. 4.13) frequently referred to as the constant-breadth cam is the simplest positive-drive type. The two flat follower surfaces are a fixed distance apart and contact opposite sides of the cam. The follower may either translate or oscillate. In contrast with other cams, this constant breadth is a restriction which allows only 180 degrees of cam action to be specified. The other 180 degrees of action must be a mirror image so as to maintain a constant width d.

Constant-breadth harmonic (circular arc) cams have been success-fully used for small mechanisms, such as sewing machine feeds.

In the example that follows (Fig. 4.13), we are given the follower width d and the displacement diagram with a rise of h in. in 150 degrees

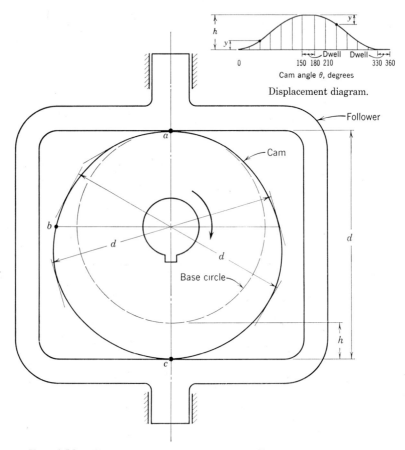

Fig. 4.13.   Yoke radial cam—flat-faced follower (positive drive).

of cam rotation and 30 degrees of dwell. The layout procedure is as follows:

(a) Duplicating the steps of Art. 4.4, construct the cam profile for 180 degrees of cam action, giving contour a, b, c.

(b) Locate points for the other half of the cam by drawing flat sides a distance d diametrically opposite the established cam profile surface a, b, c. Referring to the displacement diagram, we see that the return action is a mirror image of the rise.

Note that distance d equals the base-circle diameter plus the rise of the follower h.

4.12 YOKE RÁDIAL CAM—OSCILLATING, DUAL ROLLER FOLLOWER (POSITIVE DRIVE)

Roller followers may be utilized as an improvement over the flat-faced followers of the previous article. However, these roller followers do not produce an exact constant-breadth cam, since the cam surface is always tangent to the roller. The cam layout procedure is similar to

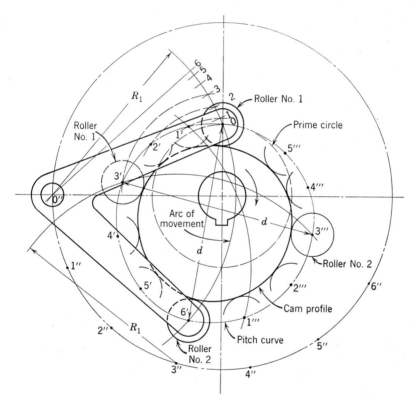

Fig. 4.14. Yoke radial cam—oscillating dual roller follower (positive drive).

the methods shown previously, although now the trace points are maintained a constant distance d apart. In the example, we shall use an oscillating roller follower. It will be seen that the follower action is controlled for 180 degrees of cam rotation minus a small follower oscillating angle.

We are given (Fig. 4.14) the oscillating roller follower no. 1, the prime-circle diameter, pivot location 0″, and the arc of movement. The follower movement shown on arc 0, 1, 2, 3, etc., is to occur while the cam rotates 150 degrees. The 6 divisions of arc are for equal cam rotation angles of 25 degrees each. The complementary roller no. 2 is to be located on the arc of movement and on the cam side opposite the other roller.

Let us construct the cam:

(a) Draw the circle, locating the follower pivot positions at 6 points, 0″, 1″, 2″, 3″, etc., at 25-degree intervals, totaling 150 degrees.

(b) From these points, swing arcs R_1 in. long, giving trace points 1′, 2′, 3′, etc., intersecting the arc of its respective rise point. Position 3 showing point 3′ is indicated.

(c) From these trace points, swing an arc d in. long, intersecting arc R_1, giving the trace points 1‴, 2‴, 3‴, etc., of roller no. 2. Point 3‴ is shown.

(d) The pitch curve, rollers, and tangents are drawn constructing the cam profile.

4.13 CONJUGATE YOKE RADIAL CAM—TRANSLATING, DUAL ROLLER FOLLOWER (POSITIVE DRIVE)

This cam generally provides the best performance of the positive-drive types. It consists of two radial cams, respectively, in contact with two conjugate (translating or oscillating) roller followers placed on opposite sides of the cam-shaft. One of the cams is for the rise period of the follower action, and the other for the fall period. This cam mechanism does not have the backlash or wear problems of the roller-groove or constant-breadth types, nor the limited 180-degree control of the latter. It is exact and dependable under high speeds or high loads. However, this cam type generally requires a larger mechanism and is slightly more expensive than the others. Nevertheless, its dependability and action very often justify its choice. In the author's opinion, the action of many existing designs could be improved by its application.

In the example (Fig. 4.15), we shall use a translating follower having rollers on diametrically opposite sides of the cam. The displacement diagram, equal prime-circle diameters, and roller-follower diameters are given. It should be noted that the prime circles may or may not be the same size. The layout procedure is as follows:

(a) Draw the pitch curve for points 0 through 6, using roller no. 1 on cam no. 1.

(b) Find the distance d between the trace points. For equal prime circles, d equals the prime-circle diameter plus the follower rise h.

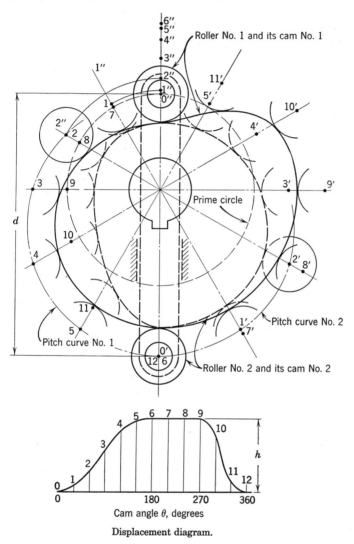

Displacement diagram.

Fig. 4.15. Conjugate yoke radial cam—translating, dual roller follower (positive drive).

(c) With distance $d = 11' = 22' = 33'$, etc., find the conjugate position of cam disk no. 2 and roller no. 2, giving pitch curve points 0', 1', 2', 3', etc.

(d) Duplicate the foregoing step (a) for the return cam, giving pitch curve points 7, 8, 9, etc., of cam disk and roller no. 2. With distance d, find trace points 7′, 8′, 9′, etc., of cam disk and roller no. 1.

(e) Draw the cam profiles tangent to the rollers.

Thus we have complete control of follower action and the conjugate rollers provide nearly perfect constraint.

4.14 CYLINDRICAL CAM

Let us indicate the steps necessary to construct a cylindrical cam. The most popular cylindrical cam is the grooved positive-drive roller type. Usually a commercially available cylindrical roller is utilized for the follower, although a conical roller is frequently chosen for longer wear life on small cylindrical cams. We shall see that the layout of the cylindrical cam is more easily accomplished than that of the radial cam. The displacement diagram is a true development of the cam surface with the pitch point and the transition point at the same place. With the displacement diagram length equal to the circumference of a cylindrical cam, the pressure angles are the same whether measured on the cam or the displacement diagram. The cylindrical cam layout may be located directly on the cylindrical surface by the application of sheet metal or paper templates. These are wrapped around the cam blank with the roller groove sides scribed on the surface. This method is, of course, inaccurate. In the following example, we shall use a cylindrical roller follower, although the procedure is the same for the conical roller type:

A positive-drive cylindrical cam is to move a translating $7/8$-in.- diameter roller follower forward and return $2\frac{1}{2}$ in. each with simple harmonic motion in 180 degrees of cam rotation. The maximum pressure angle should be approximately 30 degrees. The cam groove is $3/4$ in. deep. The contour of the cam is to be constructed. The steps are as follows:

(a) First let us find the cam size. Equation 3.19 yields a pitch-circle radius R_p of 2.28 in. Taking a fractional dimension of $2\frac{5}{16}$ in. gives the outside diameter of the cylindrical cam as

$$2\tfrac{5}{16} \times 2 + \tfrac{3}{4} = 5\tfrac{3}{8} \text{ in.}$$

The outside circumference of the cam is

$$\pi(5\tfrac{3}{8}) = 16.8 \text{ in.}$$

This is the abscissa of the development view (Fig. 4.16).

(b) In the same view, for a rise of $2\frac{1}{2}$ in., plot the SHM pitch curve, using the methods of Art. 2.9.

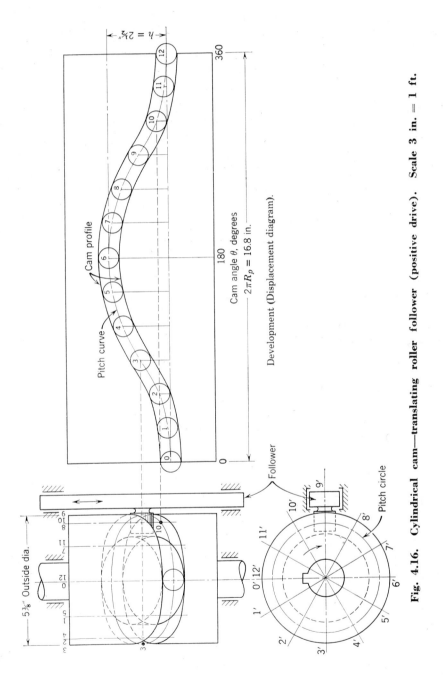

Fig. 4.16. Cylindrical cam—translating roller follower (positive drive). Scale 3 in. = 1 ft.

(c) Draw rollers with groove sides tangent. The other two views are shown for general information. They are not specifically needed in designing the cam profile.

4.15 INVERSE CAM

In all previous cams, the rotating cam profile moved at a constant speed and was the driving member that determined the action of the follower. With the inverse cam, the profile is on the follower which is driven by a constant-speed roller or pin. Star wheel or Geneva mecha-

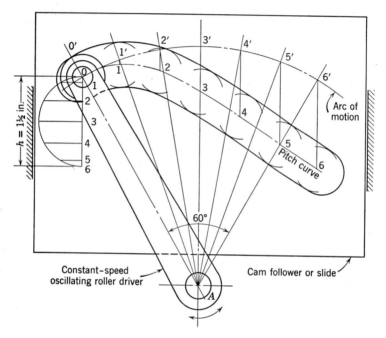

Fig. 4.17. Inverse cam—oscillating driver, translating follower. Scale ⅝″ = 1″.

nisms may be considered inverse cams. In Fig. 4.17, we see a constant-speed oscillating roller driver used to translate the follower. If a portion of the follower slot had a radius equal to the crank radius (properly located), we would have a momentary dwell. If the slot were perpendicular to the direction of the translating follower movement, we would have the familiar Scotch yoke mechanism, in which the follower action is a cosine function. It is possible to have complete rotation of the driver if the crank radius is small enough or the groove long enough. Thus, the inverse cam would have controlled action for only 180 degrees of driver rotation with the other 180 degrees a mirror image. In designing all inverse cams, one should be wary of the poor force distribution that often occurs. Frequent construction trials may be necessary until the proper contour is found. In the following example (Fig. 4.17), let us construct the contour of an inverse cam moving 1½ in. with simple harmonic motion in 60 degrees of driver rotation. The driver arm is 4 in. long. The procedure is as follows:

(a) Divide the rise, $h = 1½$ in., into its simple harmonic motion divisions, giving points 0, 1, 2, 3, etc.

(b) Divide the driver movement into the same number of equal angle displacements, giving radial lines $A1'$, $A2'$, $A3'$, etc., on the arc of motion.

(c) Locate the cam movement, e.g., length $01 = 1'1$, $02 = 22'$, $03 = 33'$, etc. The pitch curve may now be drawn.

(d) Draw rollers with the cam groove sides tangent. Note: the return action is a mirror image of the rise.

Calculating Cam Profiles

4.16 INTRODUCTION

As cam speeds increase the foregoing layout method often becomes inadequate for establishing accurate cam dimensions necessary in fabrication. Small errors in the profile are found to increase the dynamic loads and vibratory amplitudes considerably. Thus, a more precise determination of profiles is required. By calculating the location of the cutter or grinder used to form the cam, a table of cam radii and corresponding cam angles is made. Sometimes it may be necessary to convert the coordinate system from polar to rectangular for higher accuracy of cam profile. The calculations should be carried out to more significant figures than is necessary in the actual application. Generally the grinder or cutter is made the same diameter as the roller follower for mathematical simplicity. The cam is then cut on a milling machine, jig borer, or other suitable machine tool, using point by point settings. Frequently the surface formed has ridges or scallops which are sometimes hand-filed or hand-ground. We shall see in Chapter 10 that the size of these ridges and the hand-operation accuracy are determined among other things by the number of increments of cutting. For accurate cams, the setting increments are sometimes as small as one-half degree.

Preparation of the aforementioned table may require the solution of as many as eight equations for each of the machine settings. These calculations, when made with equipment such as a desk calculator,

Table 4.2   Displacement Ratio y/h*

Cam Angle Divisions	Parabolic	Simple Harmonic Motion (SHM)	Cycloidal	Cam Angle Divisions	Parabolic	Simple Harmonic Motion (SHM)	Cycloidal
0	0.000 000	0.000 000	0.000 000	60	0.500 000	0.500 000	0.500 000
1	.000 139	.000 171	.000 004	61	.516 528	.513 088	.516 663
2	.000 556	.000 685	.000 031	62	.532 778	.526 168	.533 303
3	.001 250	.001 541	.000 103	63	.548 750	.539 230	.549 897
4	.002 222	.002 739	.000 243	64	.564 444	.552 264	.566 423
5	.003 472	.004 278	.000 474	65	.579 861	.565 263	.582 859
6	.005 000	.006 156	.000 818	66	.595 000	.578 227	.599 182
7	.006 806	.008 372	.001 297	67	.609 861	.591 118	.615 369
8	.008 889	.010 926	.001 933	68	.624 444	.603 956	.631 401
9	.011 250	.013 815	.002 745	69	.638 750	.616 723	.647 255
10	.013 889	.017 037	.003 756	70	.652 778	.629 410	.662 911
11	.016 806	.020 590	.004 985	71	.666 528	.642 008	.678 349
12	.020 000	.024 472	.006 451	72	.680 000	.654 509	.693 549
13	.023 472	.028 679	.008 174	73	.693 194	.666 904	.708 493
14	.027 222	.033 210	.010 171	74	.706 111	.679 184	.723 162
15	.031 250	.038 060	.012 461	75	.718 750	.691 341	.737 539
16	.035 556	.043 228	.015 058	76	.731 111	.703 368	.751 608
17	.040 139	.048 707	.017 980	77	.743 194	.715 256	.765 353
18	.045 000	.054 497	.021 241	78	.755 000	.727 005	.778 759
19	.050 139	.060 591	.024 855	79	.766 528	.738 579	.791 812
20	.055 556	.066 987	.028 835	80	.777 778	.750 000	.804 499
21	.061 250	.073 680	.033 192	81	.788 750	.761 249	.816 808
22	.067 222	.080 665	.037 938	82	.799 444	.772 320	.828 729
23	.073 472	.087 937	.043 083	83	.809 861	.783 203	.840 251
24	.080 000	.095 492	.048 635	84	.820 000	.793 893	.851 365
25	.086 806	.103 323	.054 602	85	.829 861	.804 371	.862 065
26	.093 889	.111 427	.060 990	86	.839 444	.814 660	.872 344
27	.101 250	.119 797	.067 805	87	.848 750	.824 722	.882 195
28	.108 889	.128 428	.075 050	88	.857 778	.834 565	.891 616
29	.116 806	.137 313	.082 730	89	.866 528	.844 177	.900 603
30	.125 000	.146 447	.090 845	90	.875 000	.853 553	.909 155
31	.133 472	.155 823	.099 397	91	.883 194	.862 687	.917 270
32	.142 222	.165 435	.108 384	92	.891 111	.871 572	.924 950
33	.151 250	.175 278	.117 805	93	.898 750	.880 203	.932 195
34	.160 556	.185 340	.127 656	94	.906 111	.888 573	.939 010
35	.170 139	.195 629	.137 935	95	.913 194	.896 677	.945 398
36	.180 000	.206 107	.148 635	96	.920 000	.904 508	.951 365
37	.190 139	.216 797	.159 749	97	.926 528	.912 063	.956 917
38	.200 556	.227 680	.171 271	98	.932 778	.919 335	.962 062
39	.211 250	.238 751	.183 192	99	.938 750	.926 320	.966 808
40	.222 222	.250 000	.195 501	100	.944 444	.933 013	.971 165
41	.233 472	.261 421	.208 188	101	.949 861	.939 409	.975 145
42	.245 000	.273 005	.221 241	102	.955 000	.945 503	.978 759
43	.256 806	.284 744	.234 647	103	.959 861	.951 293	.982 020
44	.268 889	.296 632	.248 392	104	.964 444	.956 772	.984 942
45	.281 250	.308 658	.262 461	105	.968 750	.961 940	.987 539
46	.293 889	.320 816	.276 838	106	.972 778	.966 789	.989 829
47	.306 806	.333 096	.291 507	107	.976 528	.971 321	.991 826
48	.320 000	.345 491	.306 451	108	.980 000	.975 528	.993 549
49	.333 472	.357 992	.321 651	109	.983 194	.979 409	.995 015
50	.347 222	.370 590	.337 089	110	.986 111	.982 963	.996 244
51	.361 250	.383 277	.352 745	111	.988 750	.986 185	.997 255
52	.375 556	.396 044	.368 599	112	.991 111	.989 074	.998 067
53	.390 139	.408 882	.384 631	113	.993 194	.991 628	.998 703
54	.405 000	.421 783	.400 818	114	.995 000	.993 844	.999 182
55	.420 139	.434 737	.417 141	115	.996 528	.995 722	.999 526
56	.435 556	.447 736	.433 577	116	.997 778	.997 261	.999 757
57	.451 250	.460 770	.450 103	117	.998 750	.998 459	.999 897
58	.467 222	.473 832	.466 697	118	.999 444	.999 315	.999 969
59	.483 472	.486 912	.483 337	119	.999 861	.999 829	.999 996
60	.500 000	.500 000	.500 000	120	1.000 000	1.000 000	1.000 000

* y = rise of cam for the respective, angle division.
h = total rise of cam for 120 divisions.

become a drudgery, are time-consuming, and are subject to error. Electronic-type calculators have been employed very satisfactorily. Time has been reduced to less than 20 percent, providing more reliable results. As cam speeds increase these newer methods of profile determination by necessity will gain in popularity.

These calculations may be made by applying the basic curve displacement formulae of Chapter 2 or the numerical data of Table 4.2. This table presents information concerning the three most popular basic curves: simple harmonic motion, parabolic, and cycloidal. It shows the rise, y, for 1 unit of total displacement h (ratio y/h) vs. divisions of total cam angle β. For convenience, these divisions are made in 120 parts being divisible by 2, 3, 4, 5, 6, 10, and 12. Obviously, it may not be necessary to utilize all 120 points in determining the cam profile. The location of the cam cutter or grinder in terms of the cam angle divisions is easily applied to milling or grinding machine setup.

4.17   RADIAL CAM—TRANSLATING, RADIAL ROLLER FOLLOWER

First, let us investigate the simplest radial cam with a radial translating follower. For every cam angle of rotation θ, we can find the follower displacement y. This gives the pitch-curve radius (Fig. 4.18)

$$r = R_a + y \qquad (4.2)$$

where R_a = radius of prime circle, in.

Example

A radial cam rotates clockwise, the follower rising 0.750 in. with cycloidal motion in 160 degrees of cam rotation. The prime circle is 3.000 in. in diameter. Find the radial distance after the cam has rotated 120 degrees.

Solution

$$\text{Division} = \tfrac{120}{160} \times 120 = 90$$

From Table 4.2, we see the ratio

$$y/h = 0.909155$$

Therefore the displacement

$$y = 0.909155(0.750) = 0.6819 \text{ in.}$$

and the radial distance from eq. 4.2 is

$$r = R_a + y = 1.500 + 0.6819 = 2.182 \text{ in.}$$

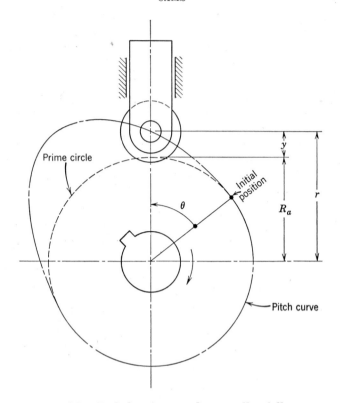

Fig. 4.18. Radial cam—translating roller follower.

4.18   RADIAL CAM—TRANSLATING, OFFSET ROLLER FOLLOWER

In this article, we shall tabulate the steps necessary to calculate the positions of the offset roller follower. This offset condition presents a more involved problem than does the radial follower previously shown. A tabulated form should be used, and the results itemized. In all previous data, we have indicated the displacement y as a selected function of cam angle θ. Now we need the values of the radius r and its respective angle ψ (Fig. 4.19). In this figure, we see that

  r = radius to trace point, in.
  y_0 = vertical displacement of prime circle, in.
  e = eccentricity or offset, in.

All other angles and distances are as shown. Thus,

$$y_0 = (R_a{}^2 - e^2)^{1/2} \tag{4.3}$$

$$r = [(y + y_0)^2 + e^2]^{1/2} \tag{4.4}$$

$$\cos (180 - \delta) = \frac{e}{r} \qquad (4.5)$$

$$\cos (180 - \Omega) = \frac{e}{R_a} \qquad (4.6)$$

$\Delta = \Omega - \delta$ } when offset opposite the (4.7)
direction of rotation, as
shown; reduced pressure
$\psi = \theta - \Delta$ } angle results (4.8)

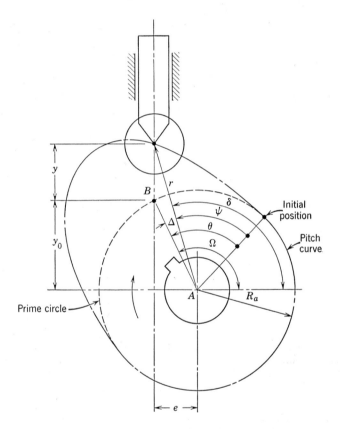

Fig. 4.19.   Radial cam—translating, offset roller follower.

Example

A radial cam turns clockwise with the follower rising 0.750 in. with cycloidal motion in 160 degrees of cam rotation. The prime circle is 3.000 in. in diameter. The offset of the follower is 0.500 in. Find the radial distance and angular roller location after the cam has rotated 120 degrees.

Solution

From eq. 4.3

$$y_0 = (R_a{}^2 - e^2)^{\frac{1}{2}} = [(1.500)^2 - (0.500)^2]^{\frac{1}{2}} = 1.414$$

From the example of Art. 4.17, the displacement

$$y = 0.682 \text{ in.}$$

From eq. 4.4,

$$r = [(y + y_0)^2 + e^2]^{\frac{1}{2}} = [(0.682 + 1.414)^2 + (0.500)^2]^{\frac{1}{2}}$$
$$= 2.155$$

From eq. 4.5,

$$\cos (180 - \delta) = \frac{e}{r} = \frac{0.500}{2.155}$$
$$\delta = 103.42 \text{ degrees}$$

From eq. 4.6,

$$\cos (180 - \Omega) = \frac{e}{R_a} = \frac{0.500}{1.500}$$
$$\Omega = 109.47 \text{ degrees}$$

From eq. 4.7,

$$\Delta = \Omega - \delta = 109.47 - 103.42 = 6.05 \text{ degrees}$$

From eq. 4.8,

$$\psi = \theta - \Delta = 120 - 6.05 = 113.95 \text{ degrees}$$

Let us tabulate the information in the accompanying table.

θ, radians	y, in.	y_0, in.	r, in.	δ, degrees	Ω, degrees	Δ, degrees	ψ, degrees
$\frac{2}{3}$	0.682	1.414	2.155	103.42	109.47	6.05	113.95

4.19 RADIAL CAM—TRANSLATING FLAT-FACED FOLLOWER

Let us find the exact profile dimensions for a radial cam in contact with a flat-faced follower. In Fig. 4.20, we see that point A is the center of the cam, AB is the line of follower motion, and point C is the instantaneous contact point between the cam and follower. As before, we are interested in finding the radius, r_c, and its respective angle, ψ_c, having been given the rise y as a function of the cam angle of rotation θ. Let

R_b = radius of base circle, in.

q = eccentricity of point of contact from cam center, in.

r_c = radial distance to cam profile, in.

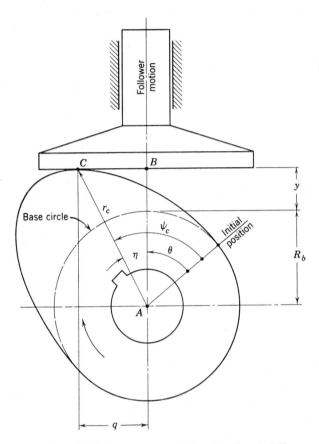

Fig. 4.20.   Radial cam—translating flat-faced follower.

We know that point C is a distance

$$q = \frac{dy}{dt} \times \frac{1}{\omega} \qquad (3.33)'$$

In Fig. 4.20, we see

$$r_c = [(R_b + y)^2 + q^2]^{1/2} \qquad (4.9)$$

$$\tan \eta = \frac{q}{R_b + y} \tag{4.10}$$

$$\psi_c = \theta + \eta \quad \text{(on the rise period of follower motion shown)} \tag{4.11}$$

$$= \theta - \eta \quad \text{(on the fall period of follower motion)} \tag{4.12}$$

The procedure for the profile determination is similar to that indicated in the previous articles. Equations 4.9 through 4.12 are applied in a tabulated form.

Furthermore, we can use some of the foregoing relationships to construct a cam for a flat-faced follower if we remember that the surface BC is always perpendicular to the line of follower motion. The steps are:

(a) For any value of cam angle θ, the distance $AB = R_b + y$.

(b) For cam angle θ, plot the value of q, obtained from eq. 3.33, perpendicular to AB from B. This gives point C which is on the cam profile. Other points may be found in the same manner.

4.20 RADIAL CAM—OSCILLATING ROLLER FOLLOWER

Now let us consider the procedure to determine the profile of a radial cam having an oscillating roller follower. The calculations will be found more cumbersome than those of the previous articles. Again, the values of radial distance r, and the respective angle ψ, must be related to known elements in the geometry of the cam in order to evaluate them for particular values of displacement y for the cam angle θ (Fig. 4.21) with all angles and distances shown. As before, we shall establish the relationships in sequence of their use. We know that the angle of oscillation ϕ is a selected function of the cam angle θ. The following equations can easily be seen from Fig. 4.21:

$$\cos \mu = \frac{l_a^2 + Q^2 - R_a^2}{2l_a Q} = \text{a constant} \tag{4.13}$$

$$\epsilon = \phi + \mu \tag{4.14}$$

$$r^2 = l_a^2 + Q^2 - 2l_a Q \cos \epsilon \tag{4.15}$$

$$\cos \sigma = \frac{r^2 + Q^2 - l_a^2}{2rQ} \tag{4.16}$$

$$\cos \lambda = \frac{Q^2 + R_a^2 - l_a^2}{2QR_a} \tag{4.17}$$

$$\Delta = \sigma - \lambda \tag{4.18}$$

$$\psi = \theta - \Delta \quad \text{(with cam rotating clockwise and} \quad (4.19)$$
$$\text{follower pivot as shown)}$$

$$= \theta + \Delta \quad \text{(with cam rotating counterclockwise} \quad (4.20)$$
$$\text{and same pivot location)}$$

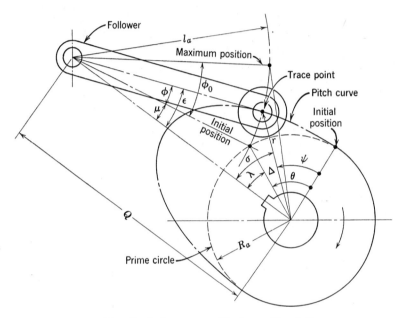

Fig. 4.21.   Radial cam—oscillating roller follower.

The accompanying table and sequence are suggested in solving the equations for the radial distance r and the cam angle ψ.

θ	ϕ	l_a	Q	R_a	μ	ϵ	r	σ	λ	Δ	ψ

4.21   RADIAL CAM—ROLLER FOLLOWER CUTTER OR GRINDER LOCATION

Conventional practice in fabricating a cam by machine tools dictates using a cutter or grinder that has the same diameter as the roller follower, if possible.   In this manner, the center of the tool will be the same place, as the follower trace point.   A cutter or grinder diameter larger than the roller results in much more involved calculations. Here the fundamental basis for the relationship is: The cutter and roller centers lie on a common normal to the cam profile. Referring to Fig. 4.22, let us establish the necessary formulae for the location of the cutter or grinder center. Let

α = pressure angle, degrees.

r_g = radial distance to center of cutter or grinder, in.

R_g = cutter or grinder radius, in.

v = follower velocity, ips.

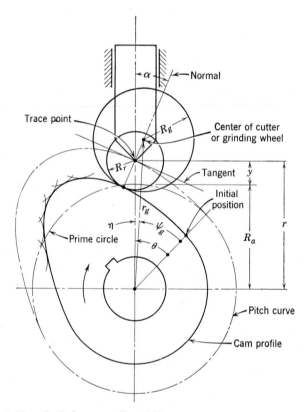

Fig. 4.22.   Radial cam roller follower cutter or grinder location.

Other angles and distances are as shown. As before, we are given the displacement y as a selected function of the cam angle θ. We need to find the radius to center of cutter r_g and its respective angle ψ_g. From Fig. 4.22, we see

$$r = R_a + y \qquad (4.21)$$

We know

$$\tan \alpha = \frac{v}{r\omega} \qquad (3.14)'$$

Again, from Fig. 4.22, we see the following

$$r_g{}^2 = r^2 + (R_g - R_r)^2 + 2r(R_g - R_r) \cos \alpha \qquad (4.22)$$

$$\cos \eta = \frac{r_g{}^2 + r^2 - (R_g - R_r)^2}{2r_g r} \qquad (4.23)$$

$$\psi_g = \theta - \eta \quad \text{(on the rise portion of the cam shown)} \qquad (4.24)$$

$$= \theta + \eta \quad \text{(on the fall portion of the cam)} \qquad (4.25)$$

The procedure for finding r_g and ψ_g is similar to that shown in the previous articles. Again, a tabulated form is suggested for ease and logical organization of the results.

4.22   RADIAL CAM—FLAT-FACED FOLLOWER CUTTER OR GRINDER LOCATION

To present the mathematical relationships between cutter or grinder and a cam that is to be used with a flat-faced follower, see Fig. 4.23. By methods of Art. 4.19, we can find the cam profile or radius r_c for every respective angle ψ_c, and also the angle η and the eccentricity, q. The following may be observed:

$$r_g = [(R_b + y + R_g)^2 + (q)^2]^{1/2} \qquad (4.26)$$

$$\cos \delta = \frac{r_g{}^2 + r_c{}^2 - R_g{}^2}{2r_g r_c} \qquad (4.27)$$

$$\psi_g = \psi_c - \delta \quad \text{(on the rise period of the follower shown)} \qquad (4.28)$$

$$= \psi_c + \delta \quad \text{(on the fall period of the follower)} \qquad (4.29)$$

Thus we can determine the cutter or grinder location (r_g and ψ_g) to produce the cam profile.

4.23   SUMMARY

In this chapter we have indicated means for the construction and calculation of the cam profile. The basis for layout is one of inversion, i.e., finding the cam profile by holding the cam stationary and moving the follower around the cam to its relative positions. We know that any curve can be used for establishing the action of the follower. However, the basic curves shown in Chapter 2 have advantages of easy layout and calculation. Also the cam size may be assumed or calculated, depending on the limiting pressure angle, cam curvature, and type of follower. As with all design problems, the final choice is a product of intelligent compromise of all the factors involved.

In terms of fabricating the cam profile, the layout method is acceptable where cam speed is limited since the contour tolerance is not better

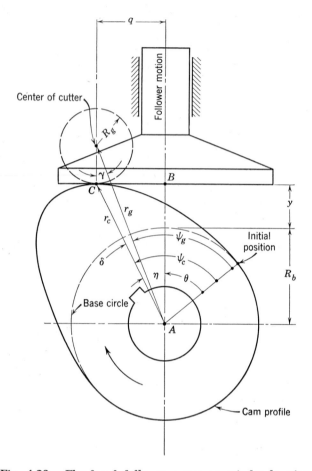

Fig. 4.23.   Flat-faced follower cutter or grinder location.

than ±.015 in. Therefore, we have presented calculations in which the profile and machine tool cutter location are determined exactly. Although this method is time-consuming, it must be employed for high-speed cams. The accuracy depends on the fabrication equipment utilized and the angular increments calculated and cut. We have shown the locations of the cutter or grinder centers in such a way that the

Legend

One required, No., A4140 steel 5/16" thick. Blank and pierce allow 3/64" on contour for finishing cam to be flat within ±0.010 after blanking. Heat-treat before machining and after blanking to Rockwell C30–34.

Induction harden cam periphery Rockwell C50–55. Master Cam No.. Tolerance on cam curve must not vary in excess of ±0.001 from true curve, profile surface 32 μ in.

Cam Profile (Roller Center)			
Points for Curve B		Points for Curve B	
Angle	Radius	Angle	Radius
0° 0' 0"	1.8345	0° 0' 0"	1.8345
0° 30' 0"	1.8345	0° 22' 30"	1.8345
1° 0' 0"	1.8345	0° 45' 2"	1.8346
1° 30' 0"	1.8346	1° 07' 38"	1.8349
2° 0' 0"	1.8349	1° 30' 18"	1.8353
2° 30' 0"	1.8353	1° 53' 5"	1.8359
3° 0' 0"	1.8359	2° 16' 1"	
Etc.	Etc.	Etc.	Etc.
23° 34' 14"	1.8388	32° 35' 28"	2.1807
24° 24' 14"	1.8369	32° 58' 8"	2.1810
24° 54' 14"	1.8355	33° 20' 44"	2.1810
25° 24' 14"	1.8347	33° 43' 15"	2.1811
25° 54' 14"	1.8345	34° 35' 46"	2.1811
150° from End of Curve A to Start of Curve B at 1.8345 Radius		150° from End of Curve B to Start of Curve A at 2.1811 Radius	

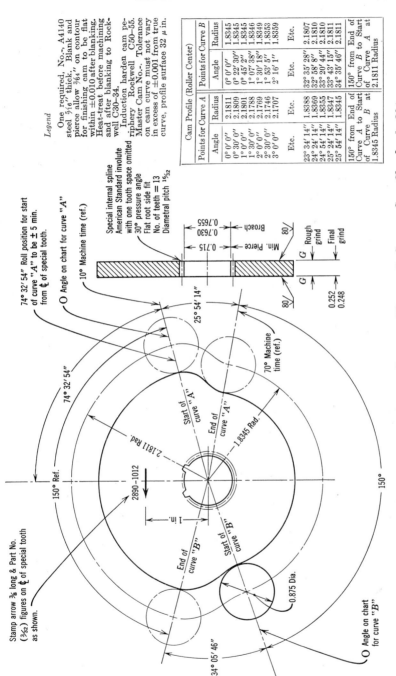

Fig. 4.24. Cam for weaving machine. Courtesy Warner & Swasey Co., Cleveland, Ohio.

information can be readily fed into the machine. A milling machine is particularly adapted, since the indexing head commonly used on these machines allows accurate setting of the angular increments. In Fig. 4.24, we see dimensions and tolerances for a high-speed weaving machine cam.

Since cams are intricate parts of mechanisms, introducing many difficulties, investigations should be made to determine whether the number of cams incorporated in a machine can be reduced. It may be possible to operate several follower levers on one cam. Also, we may have a second movement that follows a main cam most of the way and changes only a portion of its total motion. In this manner, movements may be obtained from a main cam that at first appraisal had nothing in common.

References

1. L. Kasper, *Cams, Design and Layout,* Chemical Publishing Co. Inc., New York, 1954.
2. S. Lindroth, "Calculating Cam Profiles," *Mach. Des. 23,* p. 115 (July 1951).
3. W. H. Shorter Jr., "Cam Design Analysis," *Prod. Eng. 11,* p. 223 (May 1940).
4. R. F. Griffin, "Designing Cams with Aid of Computers," *Mach. Des. 25,* p. 209 (Dec. 1953).
5. Anon., "Precision Cam Design Methods and Analysis," *Prod. Eng. 16,* p. 845 (Dec. 1945).

CHAPTER FIVE

Additional Cams and Followers

5.1

In this chapter we shall discuss the following unrelated subjects: (1) sliding and rolling actions and allied contours; (2) specially shaped radial cams; (3) cam computing mechanisms; (4) cam followers—their types and action.

These will serve to broaden the reader's theoretical and practical understanding of cams and their mechanisms.

5.2 SLIDING VELOCITIES OF BODIES IN CONTACT

We shall now analyze the general case of two sliding bodies in direct contact. This is fundamentally cam and follower action. In Fig. 5.1, body B, the driver, is turning counterclockwise, C being the driven member. At this instant, point P on body B is in contact with point Q on body C. The velocity of point P with respect to the fixed body A (absolute velocity) is equal to vector $V_{P/A}$ shown. We can resolve this absolute velocity into two perpendicular components along the tangent and the normal at the point of contact. Thus V_{PN} is the normal component of $V_{P/A}$. If these two bodies in contact are to transmit motion without crushing or separating, the velocity V_{PN} must equal the normal component of $V_{Q/A}$, V_{QN}. Therefore, the only relative motion that we may have between P and Q is along the tangent, which is the direction of sliding action. We know that $V_{Q/A}$ must be perpendicular to line 0_1Q and also have component V_{QN}. Hence it lies on the intersection of a line perpendicular to 0_1Q and the line **cb** extended. Therefore the velocity of point Q with respect to the ground A, $V_{Q/A}$ equals **Qa.**

In addition, the sliding velocity between the bodies ($V_{P/Q}$) at this

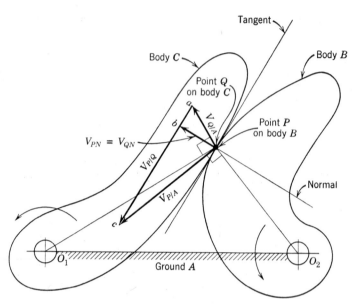

**Fig. 5.1.   Sliding velocity of bodies in contact.   At this instant $V_{P/Q}$
equals abc.**

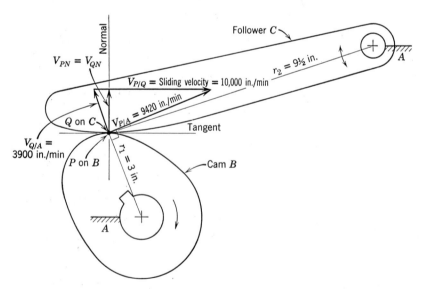

**Fig. 5.2.   Example of sliding bodies: space scale $\frac{5}{16}$ in. $= 1$ in.;
velocity scale, 1 in. $= 8000$ in./min.**

instant equals length **abc**. Therefore,

$$V_{P/Q} = V_{P/A} \to V_{Q/A} \qquad (5.1)$$

Thus, given the velocity of one body at any instant, we can find the velocity of the contacting body and the relative sliding velocity.

Furthermore, it can be shown that locking action will occur at any instant if the normal line at the contacting point passes through the center of rotation of the driven body. Excessive pressure angle is evident, and the driver will cease to transmit motion to the follower.

Example

We are given the cam and follower of Fig. 5.2 to scale. At the position shown, find the instantaneous angular velocity of the follower if the cam rotates 500 rpm.

Solution

In Fig. 5.2, draw the normal and tangent lines at the point of contact. The absolute velocity of point P on cam B,

$$V_{P/A} = r_1\omega_B = 3 \times 500 \times 2\pi = 9420 \text{ in./min}$$

Vectorially this gives the absolute velocity of point Q on follower C equal to 3900 in./min. Note that the instantaneous sliding velocity of the contacting bodies

$$V_{P/Q} = 10,000 \text{ in./min}$$

Last, we can find the instantaneous angular velocity of follower C

$$\omega_c = \frac{V_{Q/A}}{r_2} = \frac{3900}{9\frac{1}{2} \times 2\pi} = 65.3 \text{ rpm}$$

5.3 GENERAL CASE OF ROLLING CURVES

The action between two contacting bodies may be pure rolling, pure sliding, and/or a combination of both rolling and sliding. Hence, rolling surfaces are a special case of cam and follower performance.

Cams and followers usually are mixed rolling and sliding. Let us now define the term "pure rolling" as the action on a surface in which the points on the first body consecutively contact only one corresponding point on the second body. If every point on the first body comes in contact with the same point on the second body, we have "pure sliding." In Fig. 5.3, for pure rolling to occur between contacting bodies B and C, the arc QW must equal the arc PX if W meets X during the action. From Figs. 5.1 and 5.3 and eq. 5.1, we see that, for pure rolling, velocity $V_{P/Q}$ must equal zero. This occurs when velocities $V_{P/A}$ and $V_{Q/A}$ are colinear and the point of contact PQ is on the line of centers, 0_1 and 0_2. Therefore, for pure rolling action to occur between two bodies:

(*a*) the lengths of arc of action of each body must be equal;
(*b*) the point of contact between bodies must always be on the line of centers of rotation;
(*c*) the sum of the contacting distances on the line of centers of rotation must equal the distance between the centers.

Again in Fig. 5.3, let us find the contour of body B having pure rolling action with the given body C. Although the construction method shown has limited accuracy, mathematical relationships may

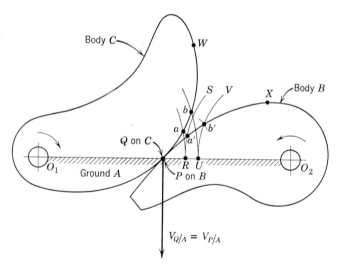

Fig. 5.3.   Pure rolling between bodies in contact.

be established for precise fabrication. Body C drives clockwise with the initial contact Q on the line of centers. Both bodies turn through a small angle. The steps for construction are as follows:

(*a*) To find point a' on body B corresponding to point a on body C, swing arc 0_1a to the line of centers, giving point R.

(*b*) With radius 0_2R, draw arc RS.

(*c*) Since arc length Pa' is equal to arc length Qa, use chord Qa as a compass distance and draw an arc from P intersecting arc RS, giving point a'.

(*d*) Find b' on C which contacts b on B as in steps (*a*) and (*b*).

Obtain U on the line of centers by swinging arc 0_1b and draw arc UV of radius 0_2U.

(e) Arc ab equals arc $a'b'$. Use chord ab as a compass distance to draw arc $a'b'$ from a', intersecting arc UV and giving point b'.

(f) Repeat to find other points.

Note that, if body B were driven in translation, we would have RS and UV as lines perpendicular to 0_10_2 since arcs RS and UV would have infinite radii. See the example of Art. 5.4 for the construction of this type. The reader should observe that construction accuracy depends upon an experienced choice of points—not too far apart nor too close together.

5.4 ROLLING BODIES OF BASIC CURVES

Pure rolling action between bodies in contact occurs with basic curves such as logarithmic spirals, ellipses, parabolas, and hyperbolas (Fig. 5.4). These curves are shown in Appendix A, which indicates both mathematical relationships and graphical layout methods. The basic proposition is:

Pure rolling action of logarithmic spirals, ellipses, parabolas, and hyperbolas requires that the centers of oscillation be located at the foci of the curves properly positioned and proportioned with respect to each other.

Figure 5.4 shows these bodies in contact at point Q, and having foci F_1 and F_2. The center of oscillation may be at either of the foci, if two are available for each curve. We see that the ellipse is the only curve that is closed and therefore the only one that will give continuous action. All other curves in contact will have limited action, i.e., oscillation. However, sectors of logarithmic spirals may be combined to give continuous rolling action. These bodies are called *lobe wheels*. For further information the reader is referred to Furman[1] and Albert and Rogers.[2]

Logarithmic spirals of equal obliquity (Figs. 5.4a and b) are shown in contact at point Q, pivoting around their foci. In Fig. 5.4a, we see the bodies rotating in opposite directions, whereas in Fig. 5.4b rotation is in the same direction. The latter is a more compact mechanism, since the action occurs on the same side of the centers of oscillation. Similar spirals are used in analog computing mechanisms.

Logarithmic spiral and translating straight-sided follower (Fig. 5.4c). It can be shown that with pure rolling the contacting curve of the straight-sided follower is a logarithmic spiral oscillating about

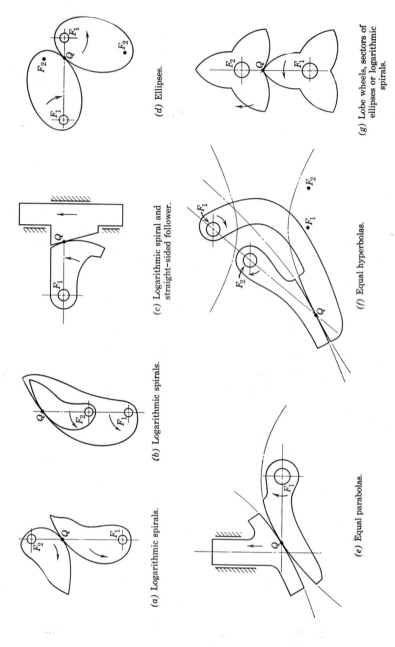

(a) Logarithmic spirals.

(b) Logarithmic spirals.

(c) Logarithmic spiral and straight-sided follower.

(d) Ellipses.

(e) Equal parabolas.

(f) Equal hyperbolas.

(g) Lobe wheels, sectors of ellipses or logarithmic spirals.

Fig. 5.4.   Pure rolling action of basic curves.

its focus. This is so because of the inherent quality of the logarithmic spiral; i.e., a tangent at all points makes a constant angle with a radial line (Appendix A). This means that the pressure angle is constant.

A *pair of equal ellipses* in Fig. 5.4d have pure rolling action which may occur for complete rotation of both bodies. For oscillation, sectors of ellipses are feasible.

A *pair of equal parabolas* is shown in Fig. 5.4e. One parabola is oscillating about its focus. The other parabolic body is translating, having the same shape as the first. The point of contact Q is on a line through the center of oscillation perpendicular to the axis of the translating body.

A *pair of equal hyperbolas* (Fig. 5.4f) may be utilized for pure rolling action, if they are properly located relative to each other. Note the location of the foci with respect to the hyperbolic curves. Hyperbolas are an excellent choice where space is at a premium.

Lobe wheels. One of the shortcomings of logarithmic spirals is that complete rotation is not possible. Figure 5.4g shows a lobe wheel made up of sectors of logarithmic spirals or ellipses which will roll with continuous action. Although tri-lobes are shown, any number may be employed.

Example

We are given a toe-and-wiper cam mechanism having a straight-sided translating follower at an acute angle of 20 degrees (Fig. 5.5). By mathematical and graphical means plot the contour of the driver having pure rolling action. Note: the contour is a logarithmic spiral, since the tangent to the curve is at a constant angle.

Solution

In Appendix A, we have the equation for the logarithmic contour

$$r = ae^{b\theta}$$

Let us find a point on the contour where $\theta = 10$ degrees. Substituting in the above,

$$r = \frac{1}{2} e^{\left(\frac{1}{\tan 20}\right)\left(\frac{10\pi}{180}\right)}$$

$$= 0.805 \text{ in. as shown}$$

The construction method is shown for the other points on the contour. These are self-explanatory and need no elaboration. We see that, the smaller the follower face angle, the larger is the size of the logarithmic cam toe necessary for the same follower displacement.

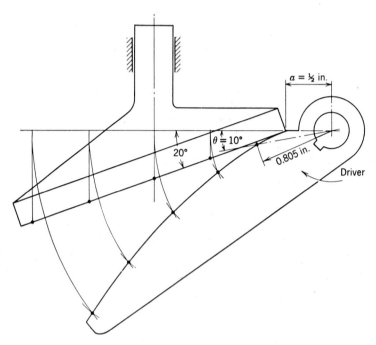

Fig. 5.5. Example of logarithmic spiral with pure rolling action. Full scale.

5.5 KINDS OF CAMS

All radial cams discussed in Chapter 4 were similar in one primary respect: the contours were developed from basic curves plotted on a displacement diagram. No attention was paid to the *actual* cam proportions. Now we shall use a reverse approach: the actual cam will be constructed of curves of various shapes, and the resulting follower movement will be determined. We know that the cam may have any shape whatsoever, as long as it fulfills the design requirements. Thus, on the radial cam, we can apply any curve or combination of lines, such as straight lines, circular arcs, parabolas, hyperbolas, ellipses, involutes,[3,6] Archimedes spirals,[9] and logarithmic spirals. However, we must remember that some of these curves have limitations because of their poor acceleration characteristics and their susceptibility to error in analysis and control. In general, these cams are not suggested for high-speed action.

For the parabolic motion and Archimedes spiral (straight-line) curves, see Chapter 2. Hyperbolic curves are not discussed, since they are rare.

In Fig. 5.6, we see some alternate combinations of curves that have been used in cam mechanisms. A circular arc dwell flank and a circular arc nose is often combined with any of the aforementioned curves. Note[4] that the three least complex conics applied to cams, such as ellipses, parabolas, and hyperbolas, have a continuous evo-

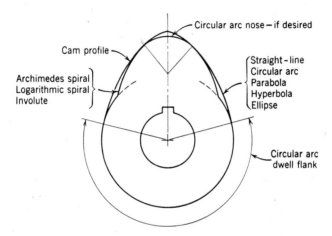

Fig. 5.6.   Radial cam composed of combinations of curves.

lute, i.e., a continuous locus of the center of curvature. This indicates continuity of the acceleration curve, assuring acceptable high-speed characteristics. However, discontinuities in evolute and acceleration curves exist when blended with circular nose and flank. In addition, the Archimedes spiral, logarithmic spiral, and involute start with an impractical abrupt slope shown. Blending curves have been employed to correct this theoretical infinite acceleration. A "triple-curve cam" having a circular arc nose, involute flanks, and a harmonic or parabolic blend into the base circle was popular in the automotive field. Of all previous curves, the circular arc is the most generally accepted. This is so because of the simplicity and accuracy in fabricating its contour in small production quantities. The cam formed solely of blending circular arcs has a discontinuous evolute.

The procedure of finding the cam shape in the past has been one of trial and error: construct the cam, plot the contour in the displacement diagram, and find the velocity and acceleration characteristics. If these need modification, change the displacement diagram to suit the design conditions and construct the actual cam. Replot if necessary until the best contour and characteristics have been obtained. Appendix A may aid the engineer in determining the basic characteristics of these curves.

5.6 LOGARITHMIC SPIRAL CAM

The logarithmic spiral has inherent qualities that make it desirable for all sorts of bodies in contact. Applied to the cam form of Fig. 5.6, it gives the smallest radial cam for a given pressure angle. Moreover, the maximum pressure angle is constant during the action as compared to the basic curves, i.e., straight-line, SHM, parabolic,

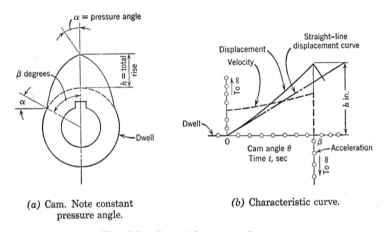

(a) Cam. Note constant pressure angle.

(b) Characteristic curve.

Fig. 5.7. Logarithmic spiral cam.

etc., which have a maximum pressure angle at one instant only. Referring to Fig. 5.7a, we see that the logarithmic curve cam has sharp corners. This gives infinite acceleration producing shock and precludes its use with a roller follower. This situation can be observed from the displacement diagram of Fig. 5.7b, which shows the logarithmic curve slope as compared to that of the basic straight-line curve. It is suggested that for proper action a blending curve at the ends of the logarithmic spiral be applied. Circular arcs and parabolic blending ramps have been successfully utilized.

Again the reader is referred to Appendix A for graphical construction and mathematical information on the logarithmic contour. For pressure angle limitation and cam size factors, the reader should refer to Fig. 3.8.

5.7 INVOLUTE CAM

The involute curve[3] is generated by a string end unwinding from a fixed circle called the involute base circle. This is not the same as the cam base circle previously defined. The involute curve when chosen for a cam has certain interesting characteristics that need elaboration. First, it can be shown that the involute curve is almost identical to the straight-line (Archimedes spiral) curve. Therefore, for a close approximation, the reader is referred to Art. 2.5, which discusses the straight-line displacement curve. These two curves

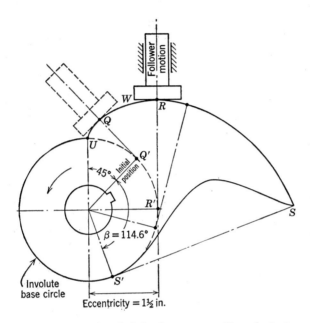

Fig. 5.8. Involute cam (single-lobe for stamp mill). Scale 6 in. = 1 ft.

have similar sharp, abrupt changes in slope at their dwell ends (Fig. 5.6). Easing-off radii may be used to permit practical application of involute curves for complete cam rotation, i.e., continuous action.

A frequent application of the involute curve is on a cam giving intermittent action to a flat-faced follower (Fig. 5.8). A cam having two or more involute lobes, called a "stamp mill cam", has been employed for pulverizing crushed ore. The inherent action is that the point of contact always occurs at the same place on the follower face. Thus, this face may be made as small as possible within practical limits. Also, if we locate the point of contact on the line of follower motion, the force components are ideal since the moments tending to bend the follower stem are minimum. Other curves require large overhangs or moments that continuously change with cam rotation. Typical is the logarithmic spiral curve of Fig. 5.5.

Another factor that gives this involute combination its excellent action is that the pressure angle equals zero at all times. Last, we shall see that cam and follower move together at a constant velocity.

Example

Show the pertinent proportions and information of one lobe of an involute stamp mill cam driving a translating flat-faced follower having a 1-in.-diameter face. The cam rotates 120 rpm. The diameter of the base circle from which the involute is generated is 3 in. The total rise of the follower is to be $h = 3$ in.

Solution

In Fig. 5.8, draw the base circle and construct the involute curve US by the method of unwinding a string from the base circle shown in Appendix A. Thus

$$\text{arc } UQ' = \text{radius } QQ'$$

and $$\text{arc } US' = \text{radius } SS'$$

Let us take the first point of cam action at point Q, which is 45 degrees from the beginning of the involute curve. This point was chosen so that the follower would not interfere with the left side of the cam body (see dotted follower). The total angle of cam rotation is

$$\beta = \frac{\text{Rise } h}{\text{Base-circle radius}} = \frac{3}{1\frac{1}{2}} \times \frac{180}{\pi}$$

$$= 114.6 \text{ degrees or 2 radians}$$

This gives points S and S'. Note: $h = SS' - QQ' = 3$ in. If the cam turns at a constant angular velocity, the follower moves at a constant speed. Let us find the follower velocity. We know the time for the cam to rotate through the total angle β

$$t = \frac{60}{120} \times \frac{2}{2\pi} = 0.159 \text{ sec}$$

The velocity of the follower

$$v = h/t = 3/0.159 = 18.9 \text{ ips} \quad (\text{constant})$$

The cam portion SS' must be designed to clear the follower on the down stroke. This is usually accomplished by making SS' a concave curve. Also we observe that the point of contact is a fixed point (on the follower face) on the eccentric line of follower motion. Action occurring beyond point S will shift the follower point of contact to the end point W, where the follower will drop downward. In the following articles, the radial translating follower will be analyzed on the basis of equivalent mechanisms. Other types of followers could be investigated in a similar manner.

5.8 ECCENTRIC CIRCLE CAM—TRANSLATING FLAT-FACED FOLLOWER

The eccentric is a type of cam. It is a circular member having its center of rotation offset from its geometric center. The follower may be of the knife-edge, flat-faced, or roller type. In this article, a translating flat-faced follower will be discussed. This combination

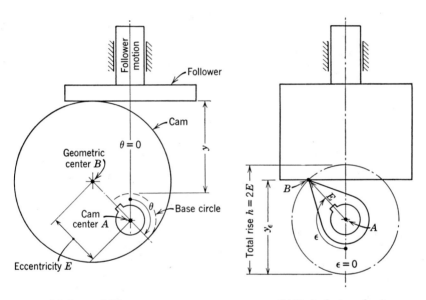

(a) Cam and follower.   (b) Equivalent mechanism.

Fig. 5.9.   Eccentric circle cam—translating flat-faced follower.
Note: ($y = y_\epsilon$ and $\theta = \epsilon$).

represents the simplest cam mechanism possible. In practice, such a mechanism is quite popular because it is simple to manufacture and has a zero pressure angle. The latter attribute almost completely eliminates the problem of jamming. In a study of cams composed of circular arcs, it is advantageous to use their equivalent linkage mechanisms. Their characteristics are easily understood and investigated. In Fig. 5.9a, we see an eccentric rotating about point A with its geometric center a distance E away. In Fig. 5.9b, we see the equivalent mechanism of the eccentric cam. This is the

Scotch yoke mechanism in which the follower movement is a simple harmonic function. Let

y = follower displacement, in.

y_ϵ = equivalent mechanism follower displacement, in.

θ = cam angle rotation for displacement y, radians.

ϵ = crank angle rotation (equivalent mechanism) for displacement y_ϵ, radians.

E = distance from cam center to circular-arc center of curvature, in.

$h = 2E$ = maximum displacement of follower, in.

ω = cam and equivalent mechanism angular velocity, radians/sec.

In this example, $y = y_\epsilon$ and $\theta = \epsilon$. From Fig. 5.9b, we see that the follower displacement

$$y_\epsilon = E(1 - \cos \epsilon) \quad \text{in.} \tag{5.2}$$

Differentiating, we find the velocity and acceleration

$$v = E\omega \sin \epsilon \quad \text{ips} \tag{5.3}$$

$$a = E\omega^2 \cos \epsilon \quad \text{in./sec}^2 \tag{5.4}$$

We see that the eccentric circle size has no effect on the follower action; only the eccentricity E does. Furthermore, offsetting the line of follower motion from the cam center of rotation does not change the follower movement.

5.9   ECCENTRIC CIRCLE CAM—TRANSLATING ROLLER FOLLOWER

Now let us use the eccentric circle cam with a translating roller follower. In Fig. 5.10a, we see an eccentric rotating about its center A, offset by a distance E. Figure 5.10b shows the equivalent mechanism, the slider crank chain. The crank radius E is distance AB, and the connecting rod length M is distance BC. Since these distances are constant, the equivalent mechanism has constant proportions for the complete action of the eccentric circle cam. Let us now derive the characteristics of this cam action. Let

y_ϵ = equivalent mechanism follower displacement measured from crank-end dead center position, in.

M = equivalent mechanism connecting rod length, in.

R_r = roller follower radius, in.

τ = angle between connecting rod and follower motion, degrees.

In this example, $y = y_\epsilon$ and $\theta = \epsilon$.

From Fig. 5.10b, we see the displacement of slider C

$$y_\epsilon = E - M - E \cos \epsilon + M \cos \tau \quad \text{in.} \tag{5.5}$$

Differentiating gives velocity

$$v = E\omega \sin \epsilon - M \sin \tau \frac{d\tau}{dt} \tag{5.6}$$

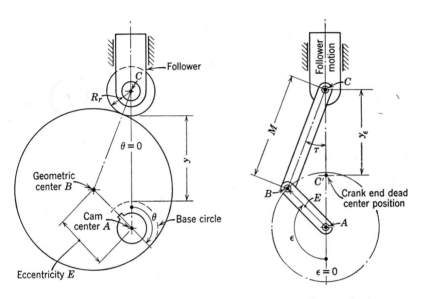

(a) Cam and follower.   (b) Equivalent mechanism.

Fig. 5.10.   Eccentric circle cam—translating roller follower.
Note: $(y = y_\epsilon$ and $\theta = \epsilon)$.

. Inspection of the figure shows the angular relationship

$$\sin \tau = \frac{E}{M} \sin \epsilon \tag{5.7}$$

Differentiating and solving yield

$$\frac{d\tau}{dt} = \frac{E\omega}{M} \frac{\cos \epsilon}{\cos \tau} \tag{5.8}$$

Substituting eq. 5.8 in eq. 5.6 gives the follower velocity, and differentiating to find the acceleration we have,

$$v = E\omega (\sin \epsilon - \cos \epsilon \tan \tau) \tag{5.9}$$

$$a = E\omega^2 \left(\cos \epsilon - \frac{E \cos^2 \epsilon}{M \cos^3 \tau} + \sin \epsilon \tan \tau \right) \tag{5.10}$$

Equations 5.5, 5.9, and 5.10 provide the characteristics of the complete follower motion. In addition, we observe that angle τ always equals the pressure angle of the cam-follower mechanism. Since we are concerned with the maximum pressure angle,

$$\tan \alpha_m = \tan \tau_m = \frac{E}{M} \tag{5.11}$$

We know that this angle has a practical limit. For more on this equivalent mechanism approach, see Spotts.[5]

Example

An eccentric circle cam drives a translating roller follower. This circle has an eccentricity of 0.500 in., a diameter of 2.500 in., and rotates at 180 rpm. The roller follower has a diameter of 0.750 in. Find (1) the characteristics of the follower motion after the cam has rotated 30 degrees from its lowest position, and (2) the maximum pressure angle.

Solution

The equivalent mechanism has a crank length E equal to 0.500 in. The connecting rod length equals the eccentric-circle radius plus the roller radius, giving

$$M = 1.250 + 0.375 = 1.625 \text{ in.}$$

The cam angular velocity

$$\omega = 180 \times \frac{2\pi}{60} = 18.85 \text{ rad/sec}$$

From eq. 5.7

$$\sin \tau = \frac{E}{M} \sin \epsilon$$

$$\tau = \sin^{-1} \left(\frac{0.500}{1.625} \sin 30 \right) = 8.85 \text{ degrees}$$

The displacement from eq. 5.5 is

$$y = E - M - E \cos \epsilon + M \cos \tau$$
$$= 0.500 - 1.625 - 0.500 \cos 30 + 1.625 \cos 8.85$$
$$= 0.0475 \text{ in.}$$

The velocity from eq. 5.9

$$v = E\omega (\sin \epsilon - \cos \epsilon \tan \tau)$$
$$= (0.500)(18.85) (\sin 30 - \cos 30 \tan 8.85) = 3.45 \text{ ips}$$

The acceleration, eq. 5.10

$$a = E\omega^2 \left(\cos \epsilon - \frac{E \cos^2 \epsilon}{M \cos^3 \tau} + \sin \epsilon \tan \tau \right)$$

$$= (0.500)(18.85)^2 \left(\cos 30 - \frac{0.500 \cos^2 30}{1.625 \cos^3 8.85} + \sin 30 \tan 8.85 \right)$$

$$= 97.5 \text{ in./sec}^2$$

The maximum pressure angle from eq. 5.11

$$\tau_m = \tan^{-1} \frac{E}{M} = \tan^{-1} \frac{0.500}{1.625} = 17.1 \text{ degrees}$$

5.10. GENERAL CASE OF CIRCULAR ARC CAMS

A cam may be composed of a combination of circular arcs (of various sizes) with or without tangent straight lines. Cams of these types have been employed very often. The primary advantages are: (a) its simplicity; (b) the low cost of manufacturing, especially in small production quantities; (c) the ease of checking for dimensional accuracy.

On the other hand, the disadvantages are the abrupt change in acceleration (infinite pulse) that occurs at the point of intersection of blending arcs or lines. This was shown in Art. 3.13 to be related to the difference between the radii of curvature at these points. High speeds are generally prohibited.

Sometimes these cams are called harmonic cams because the motion of the follower consists of parts of the simple harmonic curve. It will be shown in the following articles that each circular arc will have its respective equivalent mechanism of different proportions.

In Fig. 5.11 we have some popular combinations of shapes. Of course, any number of tangent smooth arcs and straight lines may be used. In Fig. 5.11a, we see a dwell flank arc K_3K_0 at a base-circle radius R_b, a rise arc K_0K_1 at a radius ρ_1, and an arc nose K_1K_2 having a radius ρ_n. Figure 5.11b is similar but has in addition a dwell maximum rise arc K_2K_3 at a radius ρ_2 and two equal arcs at blending radii ρ_3. Last, Fig. 5.11c shows the straight line K_0K_1 replacing the arc K_0K_1 of Fig. 5.11a. All cams except the tangent cam (Fig. 5.11c) may provide motion to either roller or flat-faced followers. Figure 5.11c is not suggested with flat-faced followers because of the high acceleration during operation: a "clicking" noise results.

Circular arcs may be substituted as an approximation for any cam shape. A trial-and-error procedure is sometimes convenient:

(a) Construct a displacement diagram, itemizing the follower movement.

(b) Plot the radial cam.

(c) Approximate this contour with arcs of a circle.

(d) Replot the displacement diagram from the circular arc cam just found.

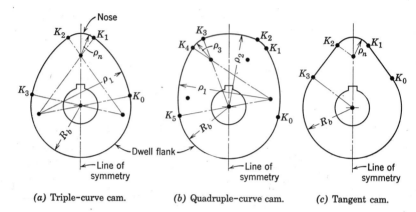

(a) Triple-curve cam.   *(b)* Quadruple-curve cam.   *(c)* Tangent cam.

Fig. 5.11. Some circular arc cams.

(e) Compare the two displacement diagrams, investigate the acceleration characteristics, and revise the cam if necessary.

In the following articles, typical examples of circular arc cams will be solved to indicate a direct approach.

5.11 CIRCULAR ARC CAM—TRANSLATING ROLLER FOLLOWER

In Fig. 5.12, we see a typical circular arc cam composed of arcs having centers at A, B_1, B_2, and B_3. Figures 5.12a and 5.12b show the cam and follower in contact over the flank arc and its equivalent slider crank mechanism. This equivalent mechanism has a crank radius E_1 equal to distance AB_3, and a connecting rod of length M_1 equal to B_3C. Equations 5.5, 5.7, 5.9, and 5.10 give the follower action. For the range of action K_0K_1, we note that the cam angle θ does not equal the equivalent mechanism angle ϵ, whereas the rise y does equal y . When the roller follower is on the cam nose (Fig. 5.12b), a different equivalent mechanism exists. This mechanism has a crank radius E_2 and a connecting rod length M_2 for the range K_1K_2. Again, the equations cited may be applied to investigate the follower characteristics. For this action, $\theta = \epsilon$ but $y \neq \epsilon$. In addition, we see that identical values of follower displacement y and velocity v exist at the blending point K_1, regardless of the equivalent mechanism. However, different acceleration values (infinite pulse) will be found; this is to be expected. The

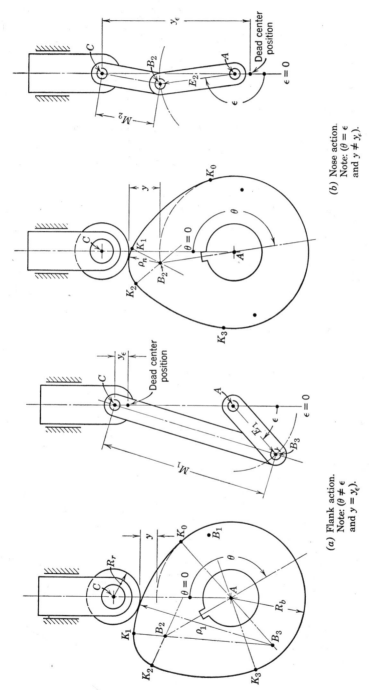

(b) Nose action.
Note: $(\theta = \epsilon$
and $y \neq y_\epsilon$).

(a) Flank action.
Note: $(\theta \neq \epsilon$
and $y = y_\epsilon$).

Fig. 5.12. Circular arc cam—translating roller follower.

reader should exercise caution in establishing the angles of rotation of the equivalent mechanism. These angles may easily be obtained from the measured layout.

Example

A cam (600 rpm) composed of circular arcs is to have a total rise of $\frac{1}{2}$ in. for a rotation of 75 degrees. The distance between follower and cam centers at maximum lift is to be $1\frac{3}{4}$ in. The roller follower is 1 in. in diameter, and the radius of the cam nose is $\frac{3}{8}$ in. Find the cam proportions, and plot curves of displacement, velocity, and acceleration.

Solution

Figure 5.13 shows the layout of the cam. We find by plotting that the base-circle radius is $\frac{3}{4}$ in. and the location of the centers of curvature are at B_1 and B_2. All proportions may be found by layout measurement or trigonometric calculation. Thus, the equivalent mechanism for action on arc K_0K_1 can be found by utilizing triangle AB_1B_2 of which AB_1 is the crank radius E_1:

$$(B_1B_2)^2 = (AB_2)^2 + E_1{}^2 - 2AB_2E_1 \cos 105$$
$$(E_1 + \tfrac{3}{4} - \tfrac{3}{8})^2 = (\tfrac{7}{8})^2 + E_1{}^2 - 2(\tfrac{7}{8})E_1 \cos 105$$
$$E_1 = 2.10 \text{ in.}$$

The flank arc radius

$$\rho_1 = 2.10 + \tfrac{3}{4} = 2.85 \text{ in.}$$

Also, the connecting rod B_1C_1

$$M_1 = 2.85 + \tfrac{1}{2} = 3.35 \text{ in.}$$

For the action over nose K_1K_2, the crank length

$$E_2 = AB_2 = \tfrac{7}{8} \text{ in.}$$

and the connecting rod length

$$M_2 = B_2C_1 = \tfrac{7}{8} \text{ in.}$$

The equivalent mechanism angles ϵ and τ may be found in a similar manner or by direct measurement. It is suggested to tabulate the data to maintain organization. Note that separate calculations for each of the two equivalent mechanisms will be made.

Substituting in eqs. 5.5, 5.7, 5.9, and 5.10 gives the values shown in Table 5.1. For greater accuracy of plotting the characteristic curves in Fig. 5.14, more points may be taken. It should be noted that the maximum value of the positive acceleration is directly dependent on the flank radius of curvature ρ_1 and the roller size R_r. Thus, the larger these radii, the greater will be the values of positive acceleration.

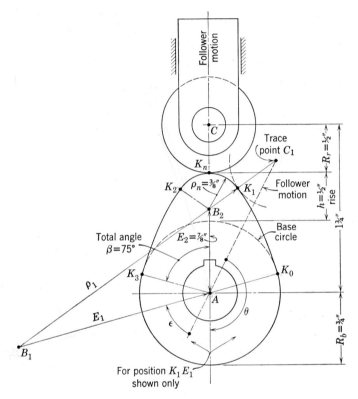

Fig. 5.13.   Circular arc cam—translating roller follower.   Full scale.

Similarly, the negative acceleration is directly dependent on the center-to-center distance E_2 and the roller size R_r. The smaller these distances, the smaller will be the negative acceleration values; this is often desirable since a smaller spring size results.

The maximum pressure angle may be checked by substituting in eq. 5.11 which indicates that this angle is not excessive. Furthermore for a more complete analysis the surface curvature and stresses should be investigated. Thus the cam and follower materials should be properly chosen to fulfill the machine functional requirements with an adequate wear life (Chapter 10).

Table 5.1   Roller Follower, Example of Art. 5.11

Point	θ, degrees	ϵ, degrees	E, in.	M, in.	τ Eq. 5.7, degrees	Sin ϵ	Cos ϵ	Tan τ	Cos τ	$E\omega$, ips	$E\omega^2$, in./sec^2	y_ϵ, Eq. 5.5, in.	y, in.	v, Eq. 5.9, ips	a, Eq. 5.10, in./sec^2
K_0	105	0	2.10	3.35	0	0	1.000	0	1.000	132.0	8250	0	0	0	3080
K_1	152	47	2.10	3.35	28	0.731	0.682	0.531	0.883	132.0	8250	0.30	0.30	50	5400
K_1	152	152	$\frac{7}{8}$	$\frac{7}{8}$	28	0.469	−0.883	0.531	0.883	55.0	3450	1.55	0.30	50	−6040
K_n	180	180	$\frac{7}{8}$	$\frac{7}{8}$	0	0	−1.000	0	1.000	55.0	3450	1.75	0.50	0	−6900

Table 5.2   Flat-Faced Follower, Example of Art. 5.12

Point	θ, degrees	ϵ, degrees	E, in.	Sin ϵ	Cos ϵ	$E\omega$, ips	$E\omega^2$, in./sec^2	y_ϵ, Eq. 5.2, in.	y, in.	v, Eq. 5.3, ips	a, Eq. 5.4, in./sec^2
K_0	105	0	2.10	0	1.000	132.0	8250	0	0	0	8250
K_1	125	20	2.10	0.342	0.940	132.0	8250	0.128	0.128	45	7720
K_1	125	125	$\frac{7}{8}$	0.820	−0.574	55.0	3450	1.38	0.128	45	−1980
K_n	180	180	$\frac{7}{8}$	0	−1.000	55.0	3450	1.75	$\frac{1}{2}$	0	−3450

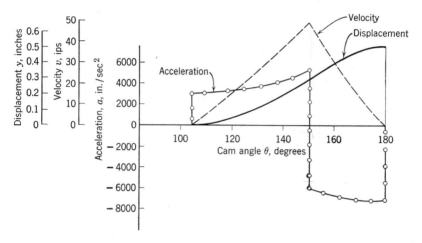

Fig. 5.14. Characteristic curves of circular arc cam—translating roller follower.

5.12 CIRCULAR ARC CAM—TRANSLATING FLAT-FACED FOLLOWER

Now we shall use the previous cam but with a flat-faced follower.[11] In Fig. 5.15, we see the Scotch yoke equivalent mechanism. As before there are two different equivalent mechanisms for each arc of contact K_0K_1 and K_1K_2. In contact on flank arc K_1K_2 (Fig. 5.15a) the center of curvature B_1 gives an equivalent mechanism, having a crank length AB_1 equal to E_1. When contact occurs on the nose arc K_1K_2, the equivalent mechanism is as shown in Fig. 5.15b. Article 5.11 indicates the method for finding the cam proportions, and eqs. 5.2, 5.3, and 5.4 should be employed to determine the follower characteristics. Figures 5.16 and 5.17 and Table 5.2 show the solution for the same cam as that given in the example of Art. 5.11 with the exception that a flat-faced follower is used in lieu of the roller type. By comparison, we observe that the characteristic curves are similar for both followers.

5.13 CIRCULAR ARC AND STRAIGHT-LINE OR TANGENT CAM—TRANSLATING ROLLER FOLLOWER

This cam[10] is often called a tangent cam since it consists of a straight line drawn tangent to circular arcs. On the cam in Fig. 5.18, we see straight sides for portions K_0K_1 and K_2K_3. Nose arc K_1K_2 is a circle with center at B_2, and arc K_0K_3 is the dwell base circle of radius R_b. The previous results of Art. 5.11 apply for the circular arc portions of the cam. However, additional equations are needed for the

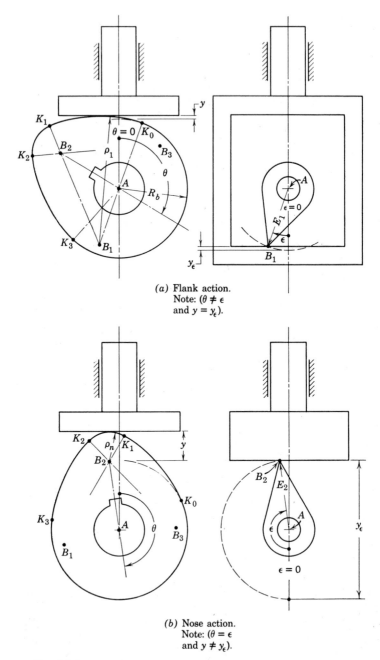

(a) Flank action.
Note: ($\theta \neq \epsilon$
and $y = y_\epsilon$).

(b) Nose action.
Note: ($\theta = \epsilon$
and $y \neq y_\epsilon$).

Fig. 5.15.   Circular arc cam—translating flat-faced follower.

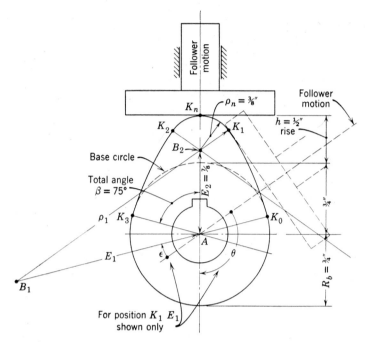

Fig. 5.16. Circular arc cam—translating flat-faced follower. Full scale.

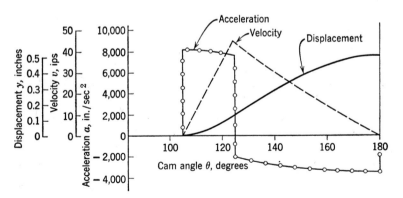

Fig. 5.17. Characteristic curves of circular arc cam—translating flat-faced follower.

straight-side portions $K_0 K_1$ and $K_2 K_3$. Let

$$\theta_t = \text{cam angle to a point on}$$
$$\text{straight-side flank, degrees.}$$

It can easily be shown that the follower displacement on straight-sided range $K_0 K_1$ or $K_2 K_3$

$$y = (R_b + R_r)\,(\sec \theta_t - 1)\quad \text{in.} \tag{5.12}$$

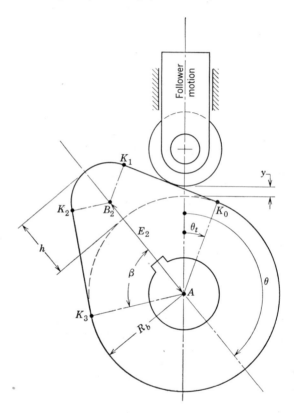

Fig. 5.18.   Tangent cam—translating roller follower.

Differentiating gives velocity and acceleration

$$v = (R_b + R_r)\omega \sec \theta_t \tan \theta_t \quad \text{ips} \tag{5.13}$$

$$a = (R_b + R_r)\omega^2(1 + 2\tan^2 \theta_t) \sec \theta_t \quad \text{in./sec}^2 \tag{5.14}$$

As before, the positive and negative accelerations depend on the cam proportions. Furthermore, the following relationships are apparent

$$R_b - \rho_n = E_2 - h = E_2 \cos \beta \tag{5.15}$$

or the total rise

$$h = E_2(1 - \cos \beta) \quad \text{in.} \tag{5.16}$$

Example

(The same problem as in the previous articles except having a straight-sided cam.)

A cam composed of a circular arc nose with straight sides is to have a maximum lift of $\frac{1}{2}$ in. for a total rotation equal to 75 degrees. The cam speed is 600 rpm. The maximum distance between cam and trace point is $1\frac{3}{4}$ in. The roller follower is 1 in. in diameter. Find the cam proportions and the follower characteristics at the intersection of the nose arc and the straight side.

Solution

We shall not show the calculations for the nose portion, since it is a repetition of the example in Art. 5.11.

From eq. 5.16, we see the distance

$$E_2 = AB_2 = \frac{h}{1 - \cos \beta}$$

$$= \frac{\frac{1}{2}}{1 - \cos 75} = 0.675 \text{ in.}$$

The cam may be constructed (Fig. 5.19) having a cam-nose radius

$$\rho_n = 1\tfrac{3}{4} - \tfrac{1}{2} - 0.675 = 0.575 \text{ in.}$$

and the base-circle radius

$$R_b = 1\tfrac{3}{4} - \tfrac{1}{2} - \tfrac{1}{2} = \tfrac{3}{4} \text{ in.}$$

The cam angular velocity

$$\omega = 600 \left(\frac{2\pi}{60}\right) = 62.8 \text{ radians/sec}$$

To find the characteristic values of point K_1 we need the cam angle θ_t at that point. By measurements or calculations, we can establish $\theta_t = 28.0$ degrees.

Substituting in eqs. 5.12, 5.13, and 5.14 gives the follower action:

$$y = (R_b + R_r)(\sec \theta_t - 1)$$
$$= (\tfrac{3}{4} + \tfrac{1}{2})(\sec 28.0 - 1) = 0.142 \text{ in.}$$
$$v = (R_b + R_r)\omega \sec \theta_t \tan \theta_t$$
$$= (\tfrac{3}{4} + \tfrac{1}{2})(62.8)(\sec 28.0 \tan 28.0) = 47.3 \text{ ips}$$
$$a = (R_b + R_r)\omega^2(1 + 2\tan^2 \theta_t) \sec \theta_t$$
$$= (\tfrac{3}{4} + \tfrac{1}{2})(62.8)^2(1 + 2\tan^2 28.0)(\sec 28.0) = 8740 \text{ in./sec}^2$$

Following the procedure of the foregoing articles, we could tabulate values and plot the follower characteristic curves, observing that the

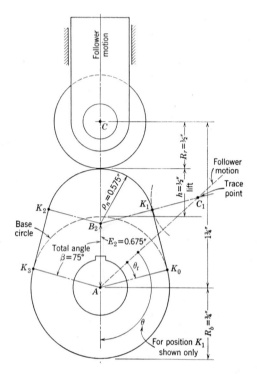

Fig. 5.19.   Tangent cam—roller follower.   Full scale.

characteristic curve shapes and values are similar to the flat-faced follower example of Art. 5.12. Furthermore the equivalent mechanism method for obtaining the cam-follower characteristics may be applied to all cams (surfaces of changing curvature) and all follower types. Analytic and graphical approaches may be utilized, and the cam curvature is approximated (by measurement) at each position under consideration. Caution is suggested in estimating this curvature. Note that a rotating cam contacting an oscillating curved-face follower has a quadric mechanism as its equivalent mechanism. The same cam mechanism (except that the follower has a flat face) may be replaced by a slider-plane equivalent mechanism. It may be mentioned that a popular graphical method employs a polar diagram in which the fixed body (having a zero acceleration) is located at the pole point.

Cam Computing Mechanisms

5.14 INTRODUCTION

Computing mechanism devices that solve problems automatically and quickly are classified as either digital or analog computers. A digital computer is one that performs the mathematical calculations with the numbers in the form of discrete valued digits. This type is capable of the basic operations of arithmetic. The most complete surveys on analog computers are by Svoboda[14] and Fry.[15] An analog device is one in which the numbers are changed into physical quantities (voltage, gear revolutions, etc.) that are related; they perform the mathematical operation which is translated into digital form. In the analog computer, the precision of the calculation depends on the precision with which the device is fabricated and read on calibrated scales. It is oftentimes a mechanical system comprised of bar linkages or cams separate or in combination, together with gears, lead screws, and other mechanical members. Applied to computers, cams are often used as correcting devices. In this manner, increased accuracy is obtained from the mechanisms. Article 11.12 shows a cam of this sort applied to the quadric mechanism.

Cams are used with mathematical relationships and functions that resist computation and because of their nature cannot be established solely by mechanical linkages. The data also may be empirical or exist in tables or curves. Cams of the mechanical type are preferred over linkages since they give more exact solutions to problems and are more straightforward to design. Although they are difficult to cut, their use for computation is often justified because of the superior accuracy attainable.* Servomechanisms added to cams amplify the small motions and small forces. In this application, the cam will perform primarily as an *information storage* device.

Cams generally fall under the category of function generators, i.e., they produce actions (usually displacements) that are definite functions of many independent variables. The first step in design is the selection of the output scale, the input scale being fixed by the fact that the cam rotates usually less than 360 degrees for the full range. The scale choice is an important consideration. From the standpoint of accuracy, it is desirable to have the scale as large as possible, i.e., as few units per inch per revolution as possible. In this manner, backlash and dimensional accuracy will have a lesser effect. Probably the most

---

* It is of interest to note that linear and non-linear potentiometers are electrical cams.

accurate cam developed was the "code-track" type utilizing sprocket holes in an input tape or other means.[26,27]

The second step is to establish the cam size and the sense of the follower movement, i.e., whether to move outward or inward for increasing values. This is chosen so that the steepest portion of the function curve is at the cam outer radius, permitting smaller cams for the same pressure angle limit. Generally a maximum pressure angle of 30 degrees is suggested but angles as high as 45 degrees have been successfully applied. In addition, the reader should beware of undercutting (see Chapter 3). This detrimental condition may be alleviated by increasing cam size or reducing the output scale. It may be mentioned that manufacturing tolerances are often by necessity better than ± 0.0003 in. and ± 3 min. Figure 10.6 shows some accurate cams.

5.15 CAM TYPES

In computers, we denote the basic function or motion relationship between non-linear related parameter X and Y as

$$Y = F(X) \qquad (5.17)$$

In this, we may call X the input parameter and Y the output parameter with $F(X)$ being any single-valued continuous function with a derivative held within certain limits. The latter restriction on $F(X)$ prevents the cam from becoming too large and impractical. Among the many functions that have been cut are squares, roots, reciprocals, trigonometric functions, and empirical functions.

In Fig. 5.20,[14] we indicate some of the more popular possibilities of cam computer mechanisms with the non-positive drive types held in contact by springs not shown. Of course, any cam may be employed. In Fig. 5.20a, we see the general case of two contours in contact usually with pure rolling action. In Fig. 5.20b, we see input X as a constant-speed rotating cam with output Y as an oscillating follower. In Fig. 5.20c, we have a spiral rotating input X with a positive drive pin in groove output Y. Figure 5.20d, often called a gear cam, is similar to Fig. 5.20c, except that the output parameter Y is a rotating pin gear meshing with grooves in the spiral input X. The primary advantage of these two spiral cams is that more than one cam revolution is allowed, providing high accuracy. Spirals as high as eight revolutions have been used. Figure 5.20e shows a rotating cylindrical cam input X and a translating output Y. Compared to a radial cam, cylindrical cams generally have a smaller maximum dimension for a given follower travel. Figure 5.20f differs from all the others in that it is a three-dimensional cam or camoid, the mechanism having 2 degrees of free-

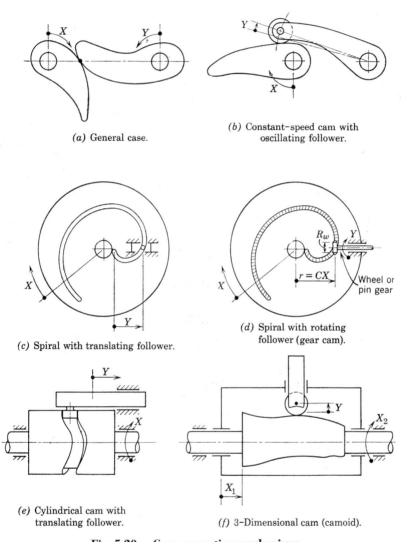

(a) General case.

(b) Constant-speed cam with oscillating follower.

(c) Spiral with translating follower.

(d) Spiral with rotating follower (gear cam).

(e) Cylindrical cam with translating follower.

(f) 3-Dimensional cam (camoid).

Fig. 5.20.   Cam computing mechanisms.

dom. The follower is often a spherical ball in contact with the cam profile. With the camoid, we have the output parameter Y a function of two parameter inputs X_1 and X_2

$$Y = F(X_1, X_2) \qquad (5.18)$$

The input X_1 and the output Y are translating with the input parameter X_2 rotating. This cam is similar to a three-dimensional graph. In an alternate design, output Y may have oscillatory action.

The three-dimensional cam is used in applications where the relation is so complicated that no analytical solution exists—the function being known only in the form of tabulated data or curves. An example is the computation of various exterior ballistic quantities such as the time of a projectile flight in fire control work. These three-dimensional cams are generally employed only as a last design possibility because of their high pressure angle, large size, high friction, and high manufacturing cost. This cam is often cut on a profiling machine controlled by a master cutter, sometimes requiring 15,000 points with a tolerance of at least $\pm$ 0.0004 in.; often higher accuracy is required. After removal from the fixture, the surface is covered with thousands of tiny protuberances between successive cuts. These projections are removed by round files properly selected to conform more or less with the cam surface at the point of filing. Last, the surface is hand-polished to 20-microinches (root-mean-square) smoothness. Sometimes to simplify fabrication these cams have been built up of many thin disks.

5.16   ARCHIMEDES SPIRAL GEAR CAM

Let us now take the Archimedes spiral as a typical example and show how a cam similar to Fig. 5.20d can be established to give the square relationship between the input and the output. This mechanism is an excellent computing tool with accuracies as high as 1 part in 30,000.

In Fig. 5.20d, we have an Archimedes spiral input parameter X, rotating clockwise, driving a pin gear parameter Y. Let

X = input cam angular rotation, radians.
Y = output follower angular rotation, radians.
R_w = mean radius of pin wheel, in.
r = instantaneous radial distance from center of cam, in.
C = a constant.

From the figure

$$R_w \, dY = r \, dX$$

But we know for the Archimedes spiral

$$r = CX$$

Therefore

$$R_w \, dY = CX \, dX$$

or

$$Y = \frac{C}{R_w} \int_0^{X_1} X \, dX$$

$$= \frac{CX^2}{2R_w} \tag{5.19}$$

since X_1 is any arbitrary end point. Therefore, the output is proportional to the square of the input cam rotation. It should be noted that offsetting the follower wheel from the cam axis would improve the force distribution of the mechanism. For the Archimedes spiral, it can be shown that an offset equal to the lead per radian will permit the follower wheel to be tangent to the spiral at all points. The form of this spiral is an involute.

The limitation of the cam discussed occurs when successive turns allow insufficient clearance for the follower. This condition can be remedied by using several cams in series—the output of the first supplying the input to the second. For example, suppose the function to be

$$Y = X^4 = [(X^2)]^2$$

This can be solved by compounding two squaring cams indicated in the equation.

5.17 BASIC SPIRAL CONTOUR CAM (NON-LINEAR FUNCTION GENERATOR)

In this article, we shall discuss the most popular cam and follower contour used in measuring, controlling, and computing equipment.[19,20,21,22] In Fig. 5.21, we see a cam of this type applied to convert a non-linear temperature function to a linear temperature reading. Linear four-bar linkages transmit the motion to and from cams which are held in contact by gravity. The basic spiral contour cam mechanism has been used in instrumentation where it was necessary to generate such quantities as trigonometric functions, ballistic functions, logarithms, probability functions, etc. For example, in multiplication the input is converted to logarithms, added, and then reconverted to a linear scale by antilogs.

This cam contour (Fig. 5.22) is the simplest, most efficient, most accurate means for linearizing input and output shafts where the motion relationship is a non-linear function. The contacting spirals which are generally used have pure rolling action and are reversible. Also, for pure rolling the contours of one curve must have an increasing radius with those of the other having a decreasing one. In other words, the curves will not have a minimum or miximum point within the range needed. The surfaces may be plain, have cross-tapes, or have gear teeth. The tape is utilized to guarantee non-slipping action between contacting surfaces. This tape is secured on each member, and transfer movement between members is by belt tension rather than friction. Gear teeth are more often employed, making the members of the non-circular gear family.

As stated previously, the given related variables are non-linear and it is necessary to make scale readings equidistant. For example, measurement of flow of a fluid may be determined by the pressure drop across an orifice. In the control field, the proportional mecha-

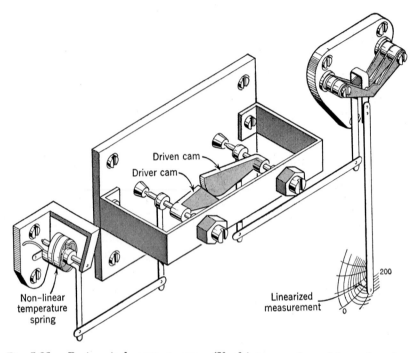

Non-linear
temperature
spring

Driven cam
Driver cam

Linearized
measurement

Fig. 5.21. Basic spiral contour cam. (Used in measuring. Linear four-bar linkages transmit motion to and from cams. Gravity holds cams in contact.)

nisms for control are linear, and therefore for best results we must again have equal scale measurement. Let us derive the basic relations for this contour. In Fig. 5.22, we see the two curves in contact. Let

θ_0 = angle rotated by the output driven member, degrees.

θ_i = the angle rotated by input driver, degrees.

ρ_0 = radius to point of contact at angle θ_0, in.

ρ_i = radius to the point of contact at angle θ_i, in.

$c_e = \rho_0 + \rho_i$ = distance between centers of rotation, in.

Both the radii and the length of periphery are important. Furthermore, as shown in Art. 5.4, for pure rolling action the point of contact P must lie on the line of centers. We know that the angle of the driven member

$$\theta_0 = f(\theta_i) \tag{5.20}$$

Differentiating,

$$\frac{d\theta_0}{d\theta_i} = f'(\theta_i) \tag{5.21}$$

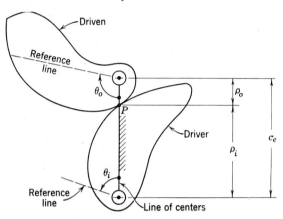

Fig. 5.22. Basic spiral contour cam.

The velocity of point P as a point on either body is equal to radius ρ multiplied by the angular velocity $d\theta/dt$ of the respective body

$$v_P = \rho_i \frac{d\theta_i}{dt} = \rho_o \frac{d\theta_o}{dt}$$

Solving,

$$\frac{d\theta_o}{d\theta_i} = \frac{\rho_i}{\rho_o} \tag{5.22}$$

Solving eqs. 5.21 and 5.22 simultaneously and knowing that $c_e = \rho_i + \rho_o$ gives radii

$$\rho_o = \frac{c_e}{1 + \dfrac{d\theta_o}{d\theta_i}} = \frac{c_e}{1 + f'(\theta_i)} \quad \text{in.} \tag{5.23}$$

$$\rho_i = \frac{c_e \dfrac{d\theta_o}{d\theta_i}}{1 + \dfrac{d\theta_o}{d\theta_i}} = \frac{c_e f'(\theta_i)}{1 + f'(\theta_i)} \quad \text{in.} \tag{5.24}$$

Equations 5.23 and 5.24 are the design equations for the contours of mating cams in terms of each radius ρ and angle θ.

It should be noted that it is not necessary that the relationships be represented by any mathematical function. The calculations can be performed by approximations from a table of numerical values. Punched card computing machines have been applied to reduce the work.

Example

Find the profile equations for the following exponential relationship where A and B are constants:

$$\theta_o = A e^{\theta_i / B}$$

Solution

We know from eq. 5.20

$$\theta_o = f(\theta_i)$$

Differentiating the given expression,

$$f'(\theta_i) = \frac{A}{B} e^{\theta_i / B}$$

Substituting in eqs. 5.23 and 5.24 gives the radii

$$\rho_o = \frac{c_e B}{B + \theta_o}$$

$$\rho_i = \frac{c_e \theta_o}{B + \theta_o}$$

Thus with the constants given tables can be established for the cam dimensions to be cut.

5.18 BASIC SPIRAL CONTOUR CAM CONSTANTS

Very often it is necessary to control the contours in which a multiple of θ_o is a function of a multiple of θ_i. Then it is necessary to compare scales of values. Let the functions

$$R = \frac{\theta_i}{K_i}, \quad S = \frac{\theta_o}{K_o} \tag{5.25}$$

Differentiating

$$\frac{d\theta_i}{dR} = K_i, \quad \frac{d\theta_o}{dS} = K_o \tag{5.26}$$

where K_i and K_o are arbitrary constants for each of the curves. Dividing

$$\frac{d\theta_o}{d\theta_i} = \frac{K_o}{K_i} \frac{dS}{dR} = \frac{K_o}{K_i} g'(R) \tag{5.27}$$

Equation 5.20 gave $\theta_o = f(\theta_i)$; it is now replaced by $S = g(R)$. Substituting eq. 5.26 in eqs. 5.23 and 5.24 gives the radii of the two contacting cams as

$$\rho_o = \frac{c_e}{1 + \dfrac{K_o}{K_i} g'(R)} \quad \text{in.} \tag{5.28}$$

$$\rho_i = \frac{c_e \dfrac{K_o}{K_i} g'(R)}{1 + \dfrac{K_o}{K_i} g'(R)} \quad \text{in.} \tag{5.29}$$

Equations 5.25, 5.28, and 5.29 yield all the information necessary to construct any cam fulfilling the relation in which $g'(R)$ does not equal zero. For information on fabrication and cutter location, the reader is referred to the excellent discussion by Hannula.[19]

Example*

It is desired to measure the temperature of a fluid A indirectly by (1) measuring the saturated vapor pressure exerted by another fluid B in a closed system subject to fluid A, and (2) then converting this pressure measurement into a linear temperature scale. The saturated vapor pressure corresponding to the temperature has been found to be of the form

$$P = 10e^{6t_m/(t_m+470)}$$

It is further assumed that this pressure can be converted to a uniform angular rotation of a shaft by means of a Bourdon tube or a similar device. The range of pressure from 10 psi to 60 psi will produce a 45.5-degree movement of the driver cam, and the full-scale range of temperatures (0 to 200 degrees F.) will be linear over a 30-degree span.

Solution

The center distance between the cams is chosen as 3 in., since this is the largest size that is practical within the space limitations. In the equation for the vapor pressure,

$$P = S \quad \text{and} \quad 10e^{6t_m/(t_m+470)} = g(R)$$

From eq. 5.26,

$$K_i = \frac{45.5}{60 - 10} = 0.91$$

* Courtesy of F. Hannula, The Foxboro Co.[19]

and

$$K_o = \frac{30}{200 - 0} = 0.15$$

From eq. 5.28

$$\rho_o = \frac{3}{1 + \dfrac{0.15}{0.91} \dfrac{d}{dt_m} 10e^{6t_m/(t_m+470)}}$$

Differentiating,

$$\rho_o = \frac{3}{1 + 16.5 \left[\dfrac{2820}{(t_m + 470)^2}\right] e^{6t_m/(t_m+470)}}$$

Also we know

$$\rho_i = c_e - \rho_o$$
$$= 3 - \rho_o \quad \text{in.}$$

From eq. 5.25,

$$\theta_i = K_i R = 0.91t_m \quad \text{degrees}$$

$$\theta_o = K_o S = K_o g(R) = 1.5e^{6t_m/(t_m+470)} \quad \text{degrees}$$

These last four equations give the information necessary for the cam construction. It is only necessary to substitute values of temperature t_m to determine the corresponding radii ρ and cam angles θ. It is conventional to use a tabular notation.

5.19 CAMS WITH SERVOMECHANISMS

Servomechanisms giving amplification of some primary motion are used widely in computing mechanisms. They take a variety of forms, including electrical, hydraulic, pneumatic, and mechanical. Servomechanisms are employed with computing cams for two reasons: (1) to strengthen the measured deviation since cams are not capable of exerting much torque or force, and (2) to correct the deviation (if any) by closed-loop or feedback schemes. The nature and profile of the cam is established by the servosystem. The cam function is primarily one of information storage and control, and not of power as in machinery application. For example, in Fig. 5.23,[25] we see a cam-controlled flame cutting mechanism. The cams contain the coordinate information for proper shaping of the workpiece. The cam followers drive X- and Y-order linear resistance potentiometers, which by closed-loop servo transmits signals to X and Y lead screws. The other equipment controls the cutting speed

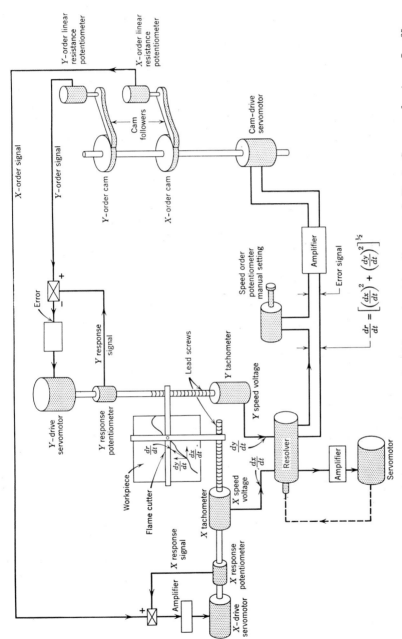

Fig. 5.23. Cam-controlled flame cutting (closed-loop circuit). Courtesy S. Davis, Servomechanisms, Inc.[25]

regulating circuit which determines the speed of the cam-drive servomotor.

Cam Followers

5.20 TYPES

Broadly speaking, there are two types of cam followers: the *roller* and the *flat-faced*. Straight cylindrical pins fixed to the follower have been employed but they are not suggested because of the excessive wear occurring on the limited area of contact. Rollers that are free to turn about an axis are chosen to distribute and reduce the wear between the cam and the follower. The roller shape may be *cylindrical, conical,* or *hyperboloidal.* Some applications having excessive deflections and misalignments of the cam or follower roller utilize a crowned roller to reduce the contact stresses. As an example, an airplane engine manufacturer uses a 6-in. radius crown on a 1¼-in.-diameter, ⅜-in.-width roller. The flat-faced follower has either a flat or a large spherical radius face. In the same manner as the previous crown, the spherical face reduces the actual surface stress due to excessive misalignment and deflection. Automobile manufacturers apply this type of follower in their valve gear mechanism. They do not utilize the roller follower because of the excessive stress in the small roller support pin.

Both the roller and flat-faced follower have merits that determine their application. The roller follower is suggested primarily where high relative velocities exist between the cam surface and its contacting body. In general, the cylindrical roller follower is the most popular because of its simplicity, ease of manufacturing, and commercial availability.

5.21 ROLLER ACTION

As we know, the roller is applied because of its low friction and low rolling wear. Pure rolling action is desired, although a tolerable amount of sliding is always evident. However, certain difficulties may arise in which the roller when running on a cam may have a considerable amount of detrimental sliding. These primary factors are: a *jammed* or *dirty bearing race, high changes* in the *cam peripheral speed* (especially in starting and stopping), *backlash,* and the *proportions of the positive-drive grooved roller follower,* if used. Let us discuss these conditions, assuming that the roller is running properly on the cam surface, i.e., with its bearing lubricated and free of dirt. It should be remembered that, although sliding is important from a wear standpoint, it may be seriously overshadowed by the indeterminate dynamic loads due to surface inaccuracies, poor profile cutting, poor cam design, and backlash in the system.

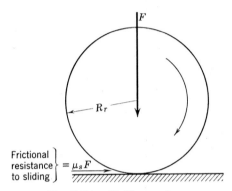

Fig. 5.24. Rolling action.

A cam rotating at a constant speed, or starting and stopping, has its peripheral speed continuously changing. This alternately accelerates and decelerates the cam roller, inducing a torque reaction which for pure rolling (Fig. 5.24) must be less than the moment of the static frictional force. Thus

$$I\alpha < \mu_s F R_r \qquad (5.30)$$

where I = moment of inertia of roller assembly, lb-in.-sec^2.

α = angular acceleration of roller, rad/sec^2.

μ_s = static coefficient of friction between cam and roller.

F = normal force, lb.

Therefore to maintain rolling and reduce sliding at high speeds a small light-weight roller is required to give the minimum inertia. The angular acceleration cannot be easily changed since it is inherent in the design. Also, the smallest cam gives the largest roller acceleration, since the ratio of roller speeds (roller acceleration) is increased for the same rise and same cam angle of rotation. Reducing sliding by increasing the coefficient of static friction or the normal force F is limited, since wear and stresses may be further increased.

5.22 GROOVE CAM ROLLER FOLLOWERS (POSITIVE DRIVE)

Followers are held in contact with the cam in one of three ways: by compression spring, by roller groove, or by opposed double rollers or surfaces. Of these, the positive roller-groove type may offer certain inherent difficulties in operation that may preclude its feasibility. The cylindrical roller in the radial or cylindrical cam groove (Fig. 5.25) must have the clearance necessary for the free movement of the roller about its axis. However, this clearance or backlash does not

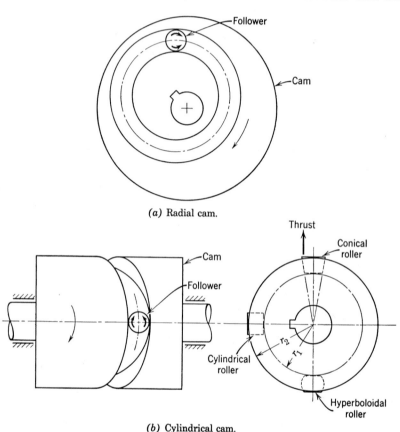

(a) Radial cam.

(b) Cylindrical cam.

Fig. 5.25. Groove cams with roller followers.

afford real constraint of the follower, since the roller will alternately contact each side of the groove, changing its direction of rotation. A kind of flutter movement sometimes occurs. With this lack of perfect constraint, both rolling and sliding actions exist for all single-roller followers in grooved cams. Cross-over shock, explained in Art. **8.9**, is related to this backlash.

For the cylindrical groove cam, another type of sliding may exist, depending on the kind of follower. As mentioned previously, the cylindrical cam has either a cylindrical, conical, or hyperboloidal follower turning about their respective axes. A cylindrical roller is preferred primarily because of the ease of cutting the groove. In the end view of Fig. 5.25b, we see a cylindrical roller in contact with surfaces between radii r_1 and r_2. This means that the velocity of the path of travel in contact with the surface at r_2 is more than the velocity of the path in contact at r_1. Sliding must occur to compensate for the velocity difference, which depends on the roller length. Obviously this length should be kept as small as possible. In general, this kind of sliding is not a seriously detrimental factor in the problem of surface life.

To improve this sliding action, a conical roller (frustrum) may be utilized. The vertex of this roller should be located at the center of the cam for best action (Fig. 5.25b). Thus, all points on the roller follower will be theoretically driven at their proper linear velocities. The conical roller has been applied largely for small cams in lieu of the simpler plain cylindrical type, because of the large ratio of radius r_1 to radius r_2 existing for the same groove depth. A special cutter must be employed to obtain the proper groove for the conical roller follower.

The conical roller in contact with a cam has a separating force component which must be overcome to keep the roller in place. However, the conical roller has the natural advantage that it can be moved radially inward to adjust for wear in the roller and groove.

Engineers have occasionally used the hyperboloidal roller follower to reduce sliding. Difficulty of manufacture and prohibitive cost have prevented its more frequent adoption. For further information on this subject, the reader is referred to Furman.[1]

In summation, we shall state that, above all other factors, clearance and backlash control the amount of sliding wear. Obviously, wear is increased by poor cam curve choice having abrupt changes in acceleration (infinite pulse).

5.23 PRACTICAL IMPROVEMENT OF POSITIVE-DRIVE ROLLERS

As stated, the backlash and sliding associated with positive-drive grooved rollers represents a primary shortcoming. Elimination of backlash is especially necessary under high-speed indeterminate dynamic and vibrational conditions. Dual rollers preloaded and in opposition present two practical avenues of improvement:[28]

(a) Two rollers in contact on either side of one groove (Fig. 5.26).

(b) A cam with rollers on both sides (Fig. 5.27).

In group (a), we may have three kinds of rollers. Figure 5.26a shows two eccentric rollers in a single groove. These rollers are of equal diameter; they are free to rotate and ride on opposite groove surfaces. An eccentric adjustment is sometimes provided for reducing backlash and preloading the roller. In Fig. 5.26b, we see concentric rollers of the same diameter but in a relieved groove. The undercut is difficult to produce accurately. The last type, and the most reasonable, is shown in Fig. 5.26c. Here we see two concentric rollers of different diameters in a stepped groove.

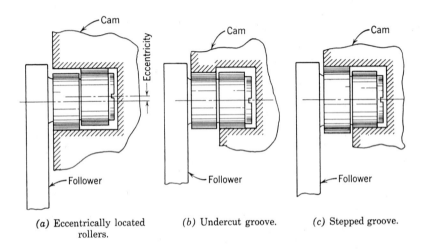

| (a) Eccentrically located rollers. | (b) Undercut groove. | (c) Stepped groove. |

Fig. 5.26.   Practical roller and groove designs.

We have three types in group (b) as well. In Figs. 5.27a and b, we have double rollers on double surfaces. Figure 5.27b is preferred because the tapered cam rib permits proper loading of the rollers, accomplished by bringing the shafts together. Most effective positive-drive action is realized by the application of conjugate or complementary cams. This utilizes dual rollers and dual cams (Fig. 5.27c). The arrangement is flexible, accurate, and reliable for high-speed, high-mass systems giving low noise and low vibrations. Preloading may be accomplished by the adjustable stud shown or by the installation of a compression spring. This method of preloading effectively controls the follower which in some installations shows an accuracy of $\pm$ 0.001 in. moving heavy follower masses weighing up to 1 ton. The author has successfully employed this type with both steel and laminated thermoset resin rollers run in oil.

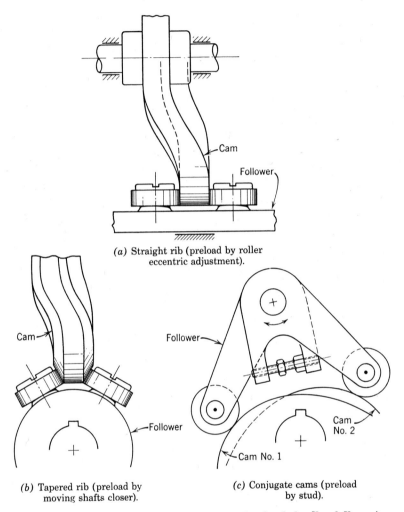

(a) Straight rib (preload by roller
eccentric adjustment).

(b) Tapered rib (preload by
moving shafts closer).

(c) Conjugate cams (preload
by stud).

Fig. 5.27. Positive-drive dual roller cams (preloaded roller follower).

5.24 ROLLER FOLLOWERS

Present-day conventional practice has established that the roller
followers be constructed of an antifriction bearing of ball, roller, or
needle construction.   Some of these are available commercially;
others are assembled from standard bearings.   The commercial follower employs a needle bearing.   Needle bearings have an advantage
over ball bearings when they are used alone as cam followers; the
large number of small-diameter, closely spaced rollers appreciably
reduce the bending stress in the outer bearing race.   The commercial

follower has an outer race (roller) of SAE 4615 or SAE 8620 carburized, hardened to Rockwell C 60–63, having a highly polished
hardened steel surface for good wear life and a tough core for

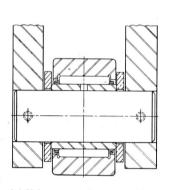

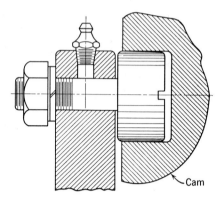

(a) Yoke-mounted commercial
follower (needle bearing
with inner race).
Courtesy of R.B.C of America.

(b) Overhang-mounted commercial follower
(needle bearing – see Table 5.3).

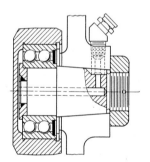

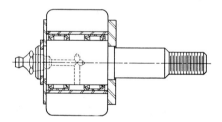

(c) Overhang-mounted follower
(double ball bearing roller
with separate outer piece
for high shock loads).[29]

(d) Overhang-mounted follower
(laminated thermoset resin
and needle bearing roller).[30]

Fig. 5.28.   Roller follower construction and installation.

absorption of dynamic and shock loads. The bearings, made of SAE
52100, are sometimes supported by a stud which is drilled to allow
for bearing lubrication. Note that journal bearings have been used.

In Fig. 5.28, we see four roller follower constructions and installations. Figure 5.28a shows a commercially available type on a simply
supported pin. This yoke mounting is preferred for open cams
because it provides greater rigidity and support for the roller pin.

However, it cannot be used for the groove or face cam, where the stud must be mounted on one side only. Figure 5.28*b* shows a cantilever-supported roller follower which may be applied in a groove cam. Obviously, with this type, much less follower support is provided, resulting in poorer stress distribution under high load. This commercial follower enjoys great popularity in general machine application, and Table 5.3 is included to show its dimensions and load capacities. The load capacity is based on the limitation of the stud strength in bending as a cantilever beam. Figure 5.28*c* shows a design used for severe shock, such as occurs with a concrete mixer roller. This roller design compensates for unavoidable shock due to pieces of stone under the rollers. The roller is a separate piece mounted on ball bearings for inexpensive replacement. It has a heavy section and is slightly softer than the bearing outer race to improve energy absorption and life of surface. Single-shielded bearings are shown with space provided that can be filled with enough high-temperature grease to last the life of the machine. Figure 5.28*d* indicates the application of needle bearings to cam rollers which have been successful in textile machinery. These rollers operate in oil at high speeds and are made of laminated thermoset resins to reduce wear and noise.

Types *c* and *d* cantilever roller followers are assembled of standard bearings. It is suggested that, when the cam is hardened to Rockwell C 60 or higher, it is not practical to exceed about one-third of the static capacity of the bearing (i.e., of the rolling elements) obtained from the commercial bearing catalogue.[29] For further discussion of cam-follower life, the reader is referred to Chapter 10.

5.25 FLAT- AND SPHERICAL-FACED FOLLOWER

A second kind of follower is the flat-faced type, which in general for the same application weighs less than the roller follower. It has the additional advantages of simplicity of maintenance and ease of lubrication. However, this follower may be applied to small cams only, since high sliding velocities are prohibitive, producing excessive stresses and short wear life due to scuffing. Obviously it cannot be employed with concave cam profiles. Lubrication with a viscous oil for this follower is essential to prevent high local temperatures and metal-to-metal contact.

Some engineers offset the mushroom follower to reduce the sliding action (Fig. 5.29). The rise or fall of the follower is unaffected by the offset. However, the follower is now rotating and translating (corkscrew action), giving sliding and rolling on the cam. The area

Table 5.3 Cam Followers (McGill Manufacturing Co.)

Dimensions, in.						Load Capacity, lb						
A	B	C	D	E	F	25 rpm	100 rpm	300 rpm	500 rpm	1000 rpm	2000 rpm	3000 rpm
1/2 *	3/8	5/8	0.190	1/4	—	490	310	210	180	140	110	100
9/16 *	3/8	5/8	0.190	1/4	—	490	310	210	180	140	110	100
5/8 *	3/8	5/8	1/4	1/4	—	730	460	320	270	210	170	150
11/16 *	7/16	3/8	1/4	5/16	1/4	730	460	320	270	210	170	150
3/4	7/16	3/8	3/8	5/16	1/4	1,670	1,050	730	610	490	390	330
7/8	7/16	3/8	3/8	5/16	1/4	1,670	1,050	730	610	490	390	330
1	1/2	1/2	3/8	1/2	1/4	2,240	1,410	980	820	650	520	450
1 1/8	1/2	1/2	7/16	1/2	1/4	2,240	1,410	980	820	650	520	450
1 1/4	5/8	5/8	7/16	5/16	5/16	4,080	2,570	1,790	1,500	1,190	950	830
1 3/8	5/8	5/8	7/16	1/2	5/16	4,080	2,570	1,790	1,500	1,190	950	830
1 1/2	3/4	3/4	1/2	3/8	3/8	5,050	3,180	2,210	1,860	1,480	1,170	1,020
1 5/8	3/4	3/4	1/2	3/8	3/8	5,050	3,180	2,210	1,860	1,480	1,170	1,020
1 3/4	7/8	7/8	1/2	3/4	7/16	6,680	4,210	2,930	2,460	1,950	1,550	1,350
1 7/8	7/8	7/8	1/2	3/4	7/16	6,680	4,210	2,930	2,460	1,950	1,550	1,350
2	1	1	1/2	1	1/2	8,590	5,410	3,760	3,160	2,510	1,990	1,740
2 1/4	1	1	1	1	9/16	8,590	5,410	3,760	3,160	2,510	1,990	1,740
2 1/2	1 1/4	1 1/4	1	1 1/8	9/16	12,540	7,900	5,500	4,620	3,670	2,910	
2 3/4	1 1/4	1 1/4	1 1/4	1 1/8	5/16	12,540	7,900	5,500	4,620	3,670	2,910	
3	1 1/2	1 1/2	1 1/4	1 1/4	5/8	17,040	10,740	7,470	6,280	4,990		
3 1/4	1 1/2	1 1/2	1 1/4	1 1/4	5/8	17,040	10,740	7,470	6,280	4,990		
3 1/2	1 3/4	2 1/4	1 3/8	1 3/8	1 1/16	23,360	14,720	10,240	8,610	6,840		
4	2 1/4	3 1/2	1 1/2	1 1/2	1 3/4	32,500	20,480	14,250	11,980	9,520		

* This bearing does not have a radial oil hole.

Notes –

1. The bearing may be lubricated from either end of the stud or from the radial hole.

2. Load capacities at 25 rpm allow a safety factor of approximately 5 when compared with the ultimate shear strength of the cam follower studs.

3. All load capacities based on an average life expectancy of 2500 hours.

4. Usually the cam-follower capacity exceeds that of the cam, unless the cam surface is hardened. Under heavy load, use large cam-followers to reduce cam surface contact stress.

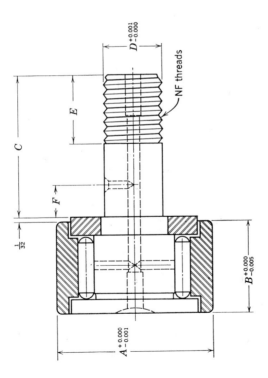

of contact and wear life is thus increased. The magnitude of the unbalanced forces on the follower stem, tending to deflect and jam it in its guide, limits the amount of offset. The practical net improvement of offsetting in this manner is subject to controversy. The author finds a limited practical overall advantage of increased wear life due to offsetting.

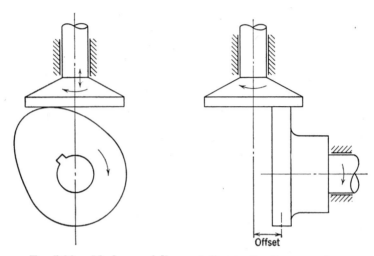

Fig. 5.29.   Mushroom follower (offset to distribute wear).

For reduced stress, some manufacturers suggest a large spherical radius (30 to 300 in.) on the face of the follower. Often designs would be improved, if the roller follower were replaced by the flat-faced follower of properly chosen materials.

A point of interest is that some manufacturers are experimenting with hyperbolic follower guides instead of cylindrical guides, the advantages being an adjustment to misalignment and deflection of cam and therefore lower surface stresses. However, high manufacturing costs are likely to preclude its adoption in the immediate future.

References

1. F. DeR. Furman, *Cams, Elementary and Advanced,* John Wiley & Sons, New York, 1921.
2. D. C. Albert and F. S. Rogers, *Kinematics of Machinery,* First Ed., John Wiley & Sons, New York, 1931.
3. F. W. Shaw, "Equations for the Design of Involute Cams," *Prod. Eng. 4,* p. 131 (April 1933).

4. R. V. Hopkins, "Cam Profiles for Minimum Inertia Shocks," *Prod. Eng. 4*, p. 174 (May 1933).
5. M. F. Spotts, "Straight-Line Follower Motions Obtained with Circular Arc Cams," *Prod. Eng. 21*, p. 110 (Aug. 1950).
6. A. H. Candee, "Formulas for Involute Curve Layouts," *Prod. Eng. 19*, p. 145 (Aug. 1948).
7. W. R. Thomson, *The Simple Cams—Having Profiles of Circular Arcs and Straight Lines*, First Ed., Machinery Publishing Co., Ltd., London.
8. W. H. Lee, "Cams for Operating Poppet Valves," *Mech. World 111*, p. 533 (June 12, 1942).
9. W. Richards, "Cam Design—The Constant Velocity Motion Cam," *Mach. London 48*, p. 473 (July 16, 1936).
10. W. Richards, "Cam Design—The Tangent Cam," *Mach., London 48*, p. 381 (Dec. 24, 1936).
11. W. Richards, "The Harmonic Motion Cam with Flat-footed Follower," *Mach., London 57*, p. 63 (Oct. 17, 1940).
12. G. L. Gillett, "Graphical Analysis of Circular Arc Cams," *Am. Machinist 65*, p. 671 (Oct. 21, 1926); *65*, p. 715 (Oct. 28, 1926); and *66*, p. 1012 (June 16, 1927).
13. H. Schreck, "Kinematics of Cams Calculated by Graphical Methods," *Trans. ASME 48*, p. 979 (1926).
14. A. Svoboda, *Computing Mechanisms and Linkages*, First Ed., McGraw-Hill, New York, 1948.
15. M. Fry, "Designing Computing Mechanisms, Part III, Cam Mechanisms," *Mach. Des. 17*, p. 123 (Oct. 1945).
16. A. S. Gutman, "Mechanical Linearizer," *Mach. Des. 25*, p. 175 (May 1953).
17. Engineering Research Associates, *High-Speed Computing Devices*, First Ed., McGraw-Hill, New York, 1950.
18. F. J. Murray, *The Theory of Mathematical Machines*, Kings Crown Press, New York, p. 17, 1947.
19. F. W. Hannula, "Designing Noncircular Surfaces," *Mach. Des. 23*, p. 111 (July 1951).
20. R. O. Yavne, "High Accuracy Contour Cams," *Prod. Eng. 19*, p. 134 (Aug. 1948).
21. E. Lockenvitz, J. B. Oliphint, W. C. Wilde, and J. M. Young, "Noncircular Cams and Gears," *Mach. Des. 24*, p. 141 (May 1952).
22. E. Lockenvitz, J. B. Oliphint, W. C. Wilde, and J. M. Young, "Geared to Compute," *Automation 2*, p. 37 (Aug. 1955).
23. A. H. Stillman, "Using Punched Card Equipment for Automatic Machine Tool Control," *Prod. Eng. 26*, p. 172 (June 1955).
24. F. W. Cunningham, "Controlling Machine Tools Automatically," *Mech. Eng., 76*, p. 487 (June 1954).
25. S. Davis, "Rotating Components for Automatic Control," *Prod. Eng. 24*, p. 129 (Nov. 1953).
26. R. L. Kenngott, Composite Cam and Cam Follower Mechanism, U. S. Patent 2,581,109, Jan. 1, 1952.
27. G. R. Stibitz, Function Unit, U. S. Patent 2,650,500, Sept. 1, 1953.
28. P. Grodzinski, "Improvement of Cam Mechanisms," *Mech. World 127*, p. 179 (Feb. 17, 1950).
29. T. Barish, "Ball Bearings Used As Cam Rollers," *Prod. Eng. 8*, p. 2 (Jan. 1937).
30. Catalogue, *Needle Bearings*, Ed. 32-B, The Torrington Co., Torrington, Conn.

Advanced Curves

6.1

In Chapter 2, the characteristics of the basic curves, such as the simple harmonic, parabolic, and cycloidal, were shown. These symmetrical curves are used because of their simplicity of construction and ease of analysis. However, for many machine requirements, they are inadequate. High speeds or special machine motions, such as multiple-step cams or unsymmetrical curves, may require modifications in the characteristic curves of displacement, velocity, and acceleration.

We shall consider the problem of special curve shapes in four ways. The first method is by *combining portions* of *basic curves* so that they may be used to best advantage. The second is to find the characteristic curves directly by means of *polynomial equations*. The third procedure uses increments by the *method* of *finite differences*. The fourth (less popular and not shown) approach has been utilized by the author in starting with the *fourth derivative* (d^4y/dt^4) *curve* and with trial and error combined with past experience, finding the desired cam shape. Electronic computers were employed to perform the increment integration in finding the best displacement, velocity, and acceleration curves. Although all methods are tedious and time-consuming, application of them is often necessary from the standpoint of improved follower motion.

The dwell-rise-dwell and dwell-rise-return-dwell curves will be analyzed. The rise-return-rise curves are too rare to justify special treatment. Article 2.17 discusses this type, indicating that an eccentric mechanism is best since it provides a motion curve having continuity in any number of derivatives.

Combinations of basic curves

6.2 INTRODUCTION

In the following articles, we shall present some of the more typical combinations of curves, together with pertinent discussion and analysis. These should serve as a basis for understanding the fundamental principles and methods. It is not possible to show all combinations. It will be noted that any combination of basic curves may be utilized to fulfill the requirements of the principles presented. To establish certain criteria, first the curves will be tangent at their junction point (have the same slope), thus providing a smooth, "bumpless" contour. This means continuous displacement and velocity curves. In addition, high-speed operation will be improved if the accelerations are equal at the point of intersection, i.e., finite pulse. In Chapter 8 it will be shown that vibrations in follower members are affected by the shape of the acceleration curve. Repeating:

The primary boundary conditions for combination curves are that there must be no discontinuity in the displacement or velocity curves. An advanced condition for high-speed action is that the acceleration curve must be continuous, have the lowest maximum value, and its slope (pulse) not too large.

In other words, the displacement, velocity, and acceleration values of each curve at the points of intersection should be respectively equal. Continuity of these curves is thus established.

As an aid, we have presented Table 6.1,[1] which gives a comparison of all the basic curves and some combination curves. These curves of displacement, velocity, acceleration, and pulse are all shown having a unit total displacement h in a unit angle β. Furthermore, the following boundary conditions are held:

$$y = 0 \quad \text{velocity } v = 0 \quad \text{when } \theta = 0$$

$$y = 1 \quad \text{velocity } v = 0 \quad \text{when } \theta = 1 = \beta$$

Table 6.1 should serve as a quick reference for comparative values and shapes of curves. To illustrate terminology we see that the D-R-D trapezoidal acceleration curve is a continuous function and its (derivative) pulse curve has many discontinuities. Note that maintaining continuity of the pulse curve is of little value with the existing machine tool cam profile fabricating precision. Anderson[2] shows other interesting curves.

Table 6.1   Comparison of Symmetrical Curves[1]

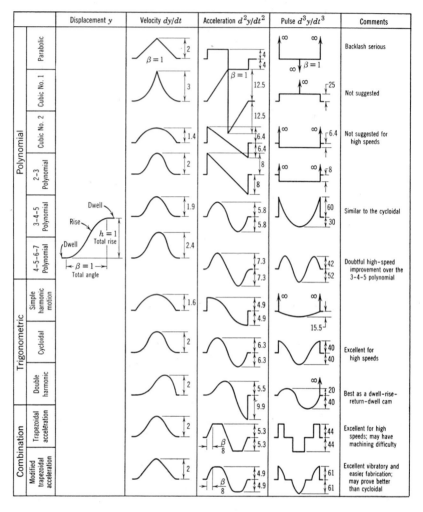

Rise or fall portions of any curve are each composed of a positive acceleration period and a negative acceleration period. For any smooth curve, we know that

$$\int_0^{\beta} a\,dt = 0 \qquad (6.1)$$

where β = cam angle rotation for total rise h, radians.

Therefore, for any rise or fall curve, the area under the positive acceleration portion equals the area under the negative acceleration portion for each curve.

This will be helpful in choosing the best acceleration curve for a cam. Therefore, any combination of basic curves is possible if the requirements, as heretofore outlined, are fulfilled. In this manner, while reducing its shortcomings, we can retain the advantages of a particular curve. With high speeds, we usually desire low accelerations and low pulse values. At other times, we are interested in controlling the ratio of the positive to the negative acceleration periods. An unsymmetrical acceleration curve, with the positive acceleration much larger than the negative acceleration (ratio about $3:1$), generally is a good choice for high-speed cams (Fig. 10.11). Smaller springs, larger cam curvature, and longer surface life result. Sometimes limited manufacturing facilities may decide the profile chosen. Many other controls may be desired to improve cam-follower action, all of which will be aided by the application of the foregoing principles in the examples that follow.

6.3   MODIFIED PARABOLIC MOTION (DWELL-RISE-DWELL CAM)

As the first example, let us combine the parabolic motion curve with the straight-line curve to give an unsymmetrical combination. In the past, this curve has been used in moderate-speed rigid linkage machines, e.g., can-making machinery. It provides fast, long-stroke action. Note that the speed is limited by the infinite pulse at the ends of the parabolic curve.

Example

A cam rotates 300 rpm, having a total (D-R-D) rise of 4 in. in 130 degrees of cam movement. The action is as follows: (1) for the first 40 degrees a positive parabolic curve acceleration; (2) for the next 30 degrees a straight-line displacement; (3) for the last 60 degrees a negative parabolic curve acceleration.

Find the ratio of accelerations, and plot all characteristic curves indicating pertinent values.

Solution

The time for a revolution $= 60/300 = 0.2$ sec/rev. In Fig. 6.1, let us divide the total action into three parts with T and Q the tangent points of the curves, giving the times

$$t_1 = 0.2 \left(\frac{40}{360} \right) = 0.0222 \text{ sec}$$

$$t_2 = 0.2 \left(\frac{30}{360} \right) = 0.0167 \text{ sec}$$

$$t_3 = 0.2 \left(\frac{60}{360} \right) = 0.0333 \text{ sec}$$

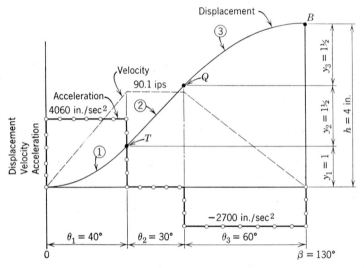

Fig. 6.1.   Parabolic straight-line combination curve
(dwell-rise-dwell cam).

We know that the displacements and the velocity for the parabolic motion of parts 1 and 3 are

$$y_1 = \tfrac{1}{2}a_1t_1^2 \quad \text{since the initial velocity } v_0 = 0 \qquad (2.33)'$$

$$y_3 = v_Qt_3 + \tfrac{1}{2}a_3t_3^2 \qquad (6.2)$$

$$v_T = a_1t_1 \qquad (6.3)$$

$$v_Q = -a_3t_3$$

For constant velocity of part 2

$$v_Q = v_T$$

Combining yields

$$a_1t_1 = -a_3t_3$$

Substituting,

$$a_1(0.0222) = -a_3(0.0333)$$

or
$$a_1 = -1\tfrac{1}{2}a_3 \qquad (6.4)$$

Also, the displacement of part 2 is

$$y_2 = v_T t_2 \qquad (6.5)$$

We are given that the total displacement

$$y_1 + y_2 + y_3 = 4$$

Substituting gives

$$\tfrac{1}{2}a_1 t_1{}^2 + v_T t_2 + v_Q t_3 + \tfrac{1}{2}a_3 t_3{}^2 = 4$$

Again substituting,

$$\tfrac{1}{2}a_1 t_1{}^2 + a_1 t_1 t_2 + a_1 t_1 t_3 - \tfrac{1}{3}a_1 t_3{}^2 = 4$$

$$\tfrac{1}{2}a_1(0.0222)^2 + a_1(0.0222)(0.0167) + a_1(0.0222)(0.0333)$$
$$- \tfrac{1}{3}a_1(0.0333)^2 = 4$$

Thus

$$a_1 = 4060 \text{ in./sec}^2$$

$$a_3 = -2700 \text{ in./sec}^2$$

The velocity at the points of intersection from eq. 6.3 is

$$v_T = v_Q = a_1 t_1 = (4060)(0.0222) = 90.1 \text{ ips}$$

The displacements for each part are

$$y_1 = \tfrac{1}{2}a_1 t_1{}^2 = \tfrac{1}{2}(4060)(0.0222)^2 = 1 \text{ in.}$$

$$y_2 = a_1 t_1 t_2 = (4060)(0.0222)(0.0167) = 1.5 \text{ in.}$$

$$y_3 = 1.5 \text{ in.}$$

We can now plot the values yielding the displacement, velocity, and acceleration curves (Fig. 6.1).

This problem may be more easily solved by using the displacement relationship $y_3 = \frac{1}{2}a_3 t_3{}^2$ in lieu of eq. 6.2, calculating from point B to Q for part 3 of the curve and using a positive sign in the final displacement formula.

6.4 TRAPEZOIDAL ACCELERATION CURVE (DWELL-RISE-DWELL CAM)

This curve[5,6] (Table 6.1) is a combination of the cubics and the parabolic curves. It modifies the parabolic curve by changing its acceleration from a rectangular to a trapezodial shape. Holst[3] and

Neklutin[7] were probably the first to appreciate that the trapezoidal acceleration curve is an improvement over the parabolic curve. If the proportions of this trapezoidal acceleration curve are reasonable, it will offer excellent vibratory characteristics under high-speed operation. To date, however, these characteristics have not proved better than those of the cycloidal curve. This trapezoidal acceleration curve has the following advantages over the basic curves:

(a) It gives a finite pulse, limited shock, wear, noise, and vibration effects as compared to the parabolic curve.

(b) It provides a smaller cam and lower peak acceleration values than the cubic no. 1 or the cycloidal curve.

(c) It offers an acceptable, smaller maximum pressure angle referred to the cubic no. 1 curve.

Let us now establish the proportions of the trapezoid with b equal to the fraction of half-rise angle $\beta/2$ at the tangent point (Fig. 6.2). It is common practice to have b between 0.2 and 0.5. If $b = 0$, we would have the rectangular parabolic motion acceleration curve, and, if $b = 0.5$, the isosceles triangular acceleration diagram (cubics) would result. A value of $b = \frac{1}{4}$ (Table 6.1) gives the most satisfactory follower performance. This curve is usually designed to be symmetrical about the transition point, although any acceleration ratio may be chosen.

6.5 MODIFIED TRAPEZOIDAL ACCELERATION CURVE (DWELL-RISE-DWELL CAM)

A combination cam that has been used in lieu of the trapezoidal acceleration curve is the modified trapezoidal type. It is composed of a parabolic motion combined with the cycloidal curve (Fig. 6.2). As before, the letter b equals the fraction of half-angle $\beta/2$ for a symmetrical curve. If $b = 0$, we have a parabolic motion curve, and, if $b = 1$, we have a cycloidal curve. Again practice shows that a value of $b = \frac{1}{4}$ gives the best overall results. This curve has the following advantages over the trapezoidal curve of the previous article:

(a) The time for the period $b \times \beta/2$ may be less for the same value of maximum acceleration, giving a smaller cam.

(b) Since the initial and final acceleration rates are higher, the cutting accuracy at these points is not as critical. This thereby reduces cost. Closer agreement with mathematical requirements is also achieved.

This cam in vibratory performance agrees closely with the trapezoidal and the cycloidal curves. The following example of a symmetrical acceleration curve is presented to indicate the mathematical fundamentals. We shall develop the curves without abrupt changes in acceleration, i.e., having continuous acceleration and pulse functions. Actually the advantage of maintaining exact mathematical continuity in the pulse curve is problematical within the present accuracies of cam

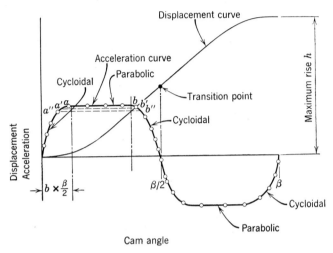

Fig. 6.2. Modified trapezoidal acceleration curve (note continuity of pulse curve).

fabrication. We may observe (Fig. 6.2) that discontinuity of the pulse curve is permissible. In other words, the intersection of two curves may be at $a'b'$ or $a''b''$ with a slight theoretical disadvantage over tangent line ab. The fabrication tolerances often are of a far greater magnitude than the displacement differences between these close curves.

Example

A cam rotates 300 rpm. A symmetrical modified trapezoidal acceleration curve (parabolic motion combined with the cycloidal curve) is to be drawn with the ratio $b = \frac{1}{4}$. The total rise is 4 in. in 160 degrees of cam rotation. Find pertinent values of all the characteristics and plot the curves.

Solution

In Fig. 6.3, we see the basic cycloidal curve from which the combination curve is developed. This figure also shows the modified trapezoidal acceleration curve. The variables pertaining to the cycloidal curve will be denoted by the primed symbols. In Fig. 6.3b, let us divide one-half of the rise into its three component parts. Since $b = \frac{1}{4}$

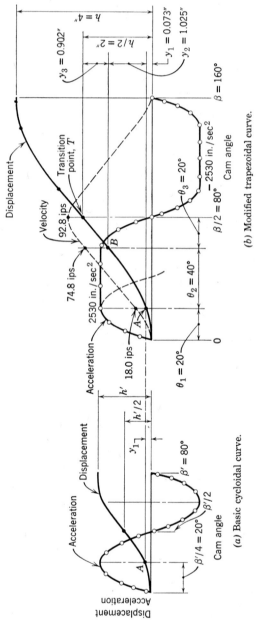

(a) Basic cycloidal curve.

(b) Modified trapezoidal curve.

Fig. 6.3. Modified trapezoidal acceleration curve (dwell-rise-dwell cam).

and the angle $\beta/2$ is 80 degrees, we see that the cam angle for parts 1 and 3 is $\theta_1 = \theta_3 = 20$ degrees $= \beta'/4$. This gives $\theta_2 = 40$ degrees. The angular velocity of the cam

$$\omega = 300/60 \times 2\pi = 31.4 \text{ rad/sec}$$

The characteristics of the cycloidal curve at point A from eqs. 2.26, 2.27, and 2.28 are

$$y_A = y_1 = \frac{h'}{4} - \frac{h'}{2\pi} = 0.091h' \tag{6.6}$$

$$v_A = \frac{h'\omega}{\beta'} = \frac{h'(31.4)}{\dfrac{80}{180}\pi} = 22.5h' \tag{6.7}$$

$$a_A = a_2 = \frac{2\pi h'\omega^2}{(\beta')^2} \tag{6.8}$$

$$= \frac{2\pi h'(31.4)^2}{\left(\dfrac{80}{180}\pi\right)^2} = 3170h'$$

The displacement of part 2, the parabolic curve, from Art. 2.12,

$$y_2 = v_A \frac{\theta_2}{\omega} + \frac{1}{2}a_2\left(\frac{\theta_2}{\omega}\right)^2 \tag{6.9}$$

Substituting, we obtain

$$y_2 = \frac{h'\omega}{\beta'}\frac{\theta_2}{\omega} + \frac{1}{2}\frac{2\pi h'\omega^2}{(\beta')^2}\left(\frac{\theta_2}{\omega}\right)^2$$

$$= \frac{h'\theta_2}{\beta'} + \frac{h'\pi\theta_2^2}{(\beta')^2} \tag{6.10}$$

$$= \frac{h'}{2} + h'\pi\left(\frac{1}{4}\right) = 1.285h'$$

Also, from eq. 2.32, the velocity at point B

$$v_B = v_A + a_2\frac{\theta_2}{\omega}$$

$$= 22.5h' + 3170h'\left(\frac{40\pi/180}{31.4}\right) = 93.3h'$$

The displacement of part 3

$$y_3 = v_B\frac{\theta_3}{\omega} + 0.091h' = 93.3h'\left(\frac{20\pi/180}{31.4}\right) + 0.091h' = 1.131h'$$

But we know that

$$y_1 + y_2 + y_3 = 2$$

Substituting,

$$0.091h' + 1.285h' + 1.131h' = 2$$

$$h' = 0.80 \text{ in.}$$

Let us now find the displacements:

$$y_1 = 0.091(0.8) = 0.073 \text{ in.}$$

$$y_2 = 1.285(0.8) = 1.025 \text{ in.}$$

$$y_3 = 1.131(0.8) = 0.902 \text{ in.}$$

Substituting to find the velocity,

$$v_A = 22.5(0.8) = 18.0 \text{ ips}$$

$$v_B = 93.3(0.8) = 74.8 \text{ ips}$$

$$v_T = \text{maximum velocity}$$

$$= v_B + {}^\Delta v_{B \text{ to } T}$$

$$= 74.8 + 18.0 = 92.8 \text{ ips}$$

Also, the maximum acceleration

$$a_A = 3170(0.8) = 2530 \text{ in./sec}^2$$

The curves may now be plotted (Fig. 6.3b). If they are compared with the trapezoidal example of the previous article, we find that this curve has a slightly lower maximum acceleration and the advantage of lower required cutting accuracy in the initial and final rise portions. Also, the vibrations induced at high speeds should be slightly smaller than those of the trapezoidal curve.

6.6 DWELL-RISE-RETURN-DWELL CAM

Now, let us consider the dwell-rise-return contour, using combination curves to improve the high-speed action. To analyze the action, we shall use the symmetrical cycloidal curve (Fig. 6.4), although any of the high-speed shapes, such as trapezoidal and modified trapezoidal, introduce the same problem. A difficulty arises that did not prevail in the dwell-rise-dwell action; i.e., an abrupt change or dip in the acceleration curve occurs at the maximum rise point. This dip is undesirable, since it produces sudden inertia loads and vibrations. It serves no purpose and should be eliminated. Therefore, the best solutions are combinations that fulfill the boundary conditions stated at the beginning of the chapter and calculated in the same manner as shown in the foregoing article.

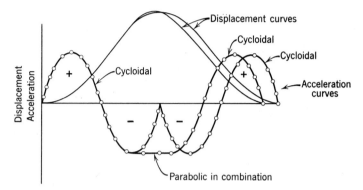

Fig. 6.4.   Dwell-rise-return-dwell curves, symmetrical rise-fall.

When we require an unsymmetrical rise-and-fall curve and an unsymmetrical positive-to-negative acceleration ratio, the mathematics becomes very involved, necessitating the solution of many simultaneous equations of boundary conditions. What we are doing is really curve fitting on a piece basis. However, a better method of solution is to employ polynomial equations, which are easier, more direct, and more accurate. They are suggested for any unsymmetrical special requirement in the dwell-rise-return-dwell curves.

Polynomial Curves

6.7

In the previous articles of this chapter, we employed the method of combining basic curves to obtain desired displacement, velocity, and acceleration curves. Now we shall utilize polynomial equations directly to find the necessary curve shapes. Stoddart[4] shows an excellent compilation of these formulae. The equations are of the form

$$y = C_0 + C_1\theta + C_2\theta^2 + C_3\theta^3 + C_4\theta^4 \cdots C_n\theta^n \qquad (6.11)$$

where y = displacement of follower, in.

θ = angle of cam rotation for follower displacement y, degrees.

C = a constant.

For dimensional notation, we shall use inches and degrees and also primes to represent the derivatives having these units. Thus

$$\text{velocity} = y' = \frac{dy}{d\theta} \quad \text{in./degree}$$

$$\text{acceleration} = y'' = \frac{d^2y}{d\theta^2} \quad \text{in./degree}^2$$

$$\text{pulse} = y''' = \frac{d^3y}{d\theta^3} \quad \text{in./degree}^3$$

By referring to Fig. 6.5, it may be seen that polynomial equations offer a difficulty when they are applied to cam design. Equation 6.11 represents only one side of the lift pattern. To find the other side, it is necessary to measure the cam angle θ plus and minus from the zero line established at the maximum rise point.

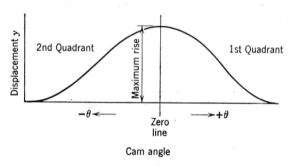

Fig. 6.5.   Polynomial equation axes.

We shall impose certain controls or limits on the end points of the cam cycle to establish the required characteristic curves. We shall see that the *boundary conditions determine the number of terms in the polynomial equation.* In addition, the higher the order of the terms, the slower will be the initial and final displacements and the more accurately the cam profile must be fabricated at the end points. Furthermore, we shall employ the polynomial method to fit an analytical expression to any given curve. That is, any curve can be closely duplicated mathematically by employing polynomial equations.

In the development that follows, a unit total rise in a unit overall time or cam angle will be assumed. All action will start in the first quadrant (Fig. 6.5), and we shall discuss the dwell-rise-dwell action first.

6.8 CONTROLLED DISPLACEMENT POLYNOMIAL (DWELL-RISE-DWELL CAM)

The simplest case of the polynomial form is the straight-line displacement curve (Fig. 6.6). The displacement eq. 6.11 becomes

$$y = C_0 + C_1\theta \qquad (6.12)$$

The boundary conditions are

$$y = 0 \quad \theta = 1, \quad \text{and} \quad y = 1 \quad \theta = 0$$

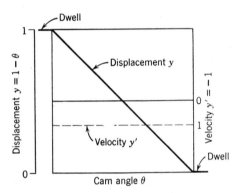

Fig. 6.6. Controlled displacement polynomial (dwell-rise-dwell cam).

Substituting in eq. 6.12 gives

$$0 = C_0 + C_1\theta$$

The displacement

$$1 = C_0$$

$$y = 1 - \theta \qquad (6.13)$$

and differentiating yields velocity

$$y' = -1 \qquad (6.14)$$

Figure 6.6 shows these relationships. Of course, this straight-line equation is impractical because of its sharp bump (theoretical infinite acceleration) at the end. The foregoing is presented primarily to indicate the method of solution.

6.9 THE 2-3 POLYNOMIAL (DWELL-RISE-DWELL CAM)

The next step in the development would be to apply controls to the velocity curve in addition to the displacement curve (Fig. 6.7). We have four boundary conditions which require a polynomial form

$$y = C_0 + C_1\theta + C_2\theta^2 + C_3\theta^3 \tag{6.15}$$

The boundary conditions are

$$\theta = 1, y = 0, y' = 0 \quad \text{and} \quad \theta = 0, y = 1, y' = 0$$

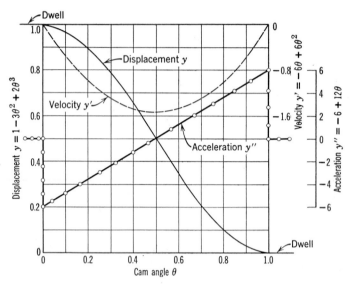

Fig. 6.7.   The 2-3 polynomial (dwell-rise-dwell cam).

Substituting in eq. 6.15 gives

$$0 = C_0 + C_1 + C_2 + C_3$$

and

$$1 = C_0$$

Differentiating eq. 6.15 yields

$$y' = 0 + C_1 + 2C_2\theta + 3C_3\theta^2$$

Substituting for velocity boundary conditions,

$$0 = 0 + C_1 + 2C_2 + 3C_3$$

and

$$0 = C_1$$

Solving these equations, the coefficients are

$$
\begin{aligned}
C_0 &= \quad 1 \\
C_1 &= \quad 0 \\
C_2 &= -3 \\
C_3 &= \quad 2
\end{aligned}
$$

giving the equations

$$y = 1 - 3\theta^2 + 2\theta^3 \tag{6.16}$$

$$y' = -6\theta + 6\theta^2 \tag{6.17}$$

$$y'' = -6 + 12\theta \tag{6.18}$$

These are plotted in Fig. 6.7. We see that, although the acceleration curve is finite at all points, the pulse is infinity at the terminal points in the curve. Shock, vibration, high stresses, and noise result at these points.

6.10 THE 3-4-5 POLYNOMIAL (DWELL-RISE-DWELL CAM)

Let us now control the contour so that its acceleration, in addition to displacement and velocity, is a finite value at all times. We have six boundary conditions which require the polynomial

$$y = C_0 + C_1\theta + C_2\theta^2 + C_3\theta^3 + C_4\theta^4 + C_5\theta^5 \tag{6.19}$$

For the boundary conditions

$$\theta = 1, \quad y = 0, \quad y' = 0, \quad y'' = 0$$

$$\theta = 0, \quad y = 1, \quad y' = 0, \quad y'' = 0$$

Differentiating eq. 6.19 gives

$$y' = C_1 + 2C_2\theta + 3C_3\theta^2 + 4C_4\theta^3 + 5C_5\theta^4$$

$$y'' = 2C_2 + 6C_3\theta + 12C_4\theta^2 + 20C_5\theta^3$$

Substituting the boundary conditions and solving the simultaneous equations yield

$$y = 1 - 10\theta^3 + 15\theta^4 - 6\theta^5 \tag{6.20}$$

$$y' = -30\theta^2 + 60\theta^3 - 30\theta^4 \tag{6.21}$$

$$y'' = -60\theta + 180\theta^2 - 120\theta^3 \tag{6.22}$$

$$y''' = -60 + 360\theta - 360\theta^2 \tag{6.23}$$

These curves are plotted in Fig. 6.8. We see that this curve family is a dynamic improvement over the 2-3 polynomial in that not only is its maximum acceleration slightly smaller but also the value of the pulse is finite at all points. Note that this 3-4-5 polynomial curve and the cycloidal curve have acceleration curves similar in shape. See Table 6.1 for comparative characteristics.

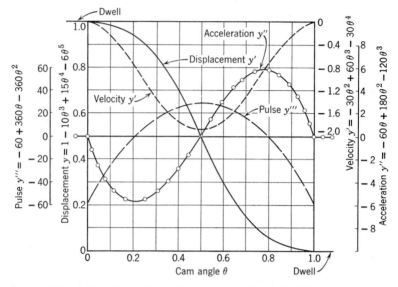

Fig. 6.8.   The 3-4-5 polynomial (dwell-rise-dwell cam).

6.11   THE 4-5-6-7 POLYNOMIAL (DWELL-RISE-DWELL CAM)

The 3-4-5 polynomial, though giving excellent high-speed character-istics, shows a discontinuity in the pulse curve at the ends. In most cams this is of little significance. However, additional controls can be introduced into the equations so that we have zero pulse at the ends. Thus, we shall control the displacement, velocity, acceleration, and pulse at the end points.

The boundary conditions require the polynomial

$$y = C_0 + C_1\theta + C_2\theta^2 + C_3\theta^3 + C_4\theta^4 + C_5\theta^4 + C_6\theta^6 + C_7\theta^7 \quad (6.24)$$

with the boundary conditions

$$\theta = 1, \quad y = 0, \quad y' = 0, \quad y'' = 0, \quad y''' = 0$$

$$\theta = 0, \quad y = 1, \quad y' = 0, \quad y'' = 0, \quad y''' = 0$$

Substituting in eq. 6.24 and its derivatives yields

$$y = 1 - 35\theta^4 + 84\theta^5 - 70\theta^6 + 20\theta^7 \quad (6.25)$$

$$y' = -140\theta^3 + 420\theta^4 - 420\theta^5 + 140\theta^6 \quad (6.26)$$

$$y'' = -420\theta^2 + 1680\theta^3 - 2100\theta^4 + 840\theta^5 \quad (6.27)$$

$$y''' = -840\theta + 5040\theta^2 - 8400\theta^3 + 4200\theta^4 \quad (6.28)$$

If we compare this acceleration curve (Fig. 6.9) with the lower order 3-4-5 polynomial, we see larger maximum acceleration and larger maximum pulse values. This indicates a possible inferiority to the 3-4-5 polynomial. The only advantage is the shift of the maximum pulse

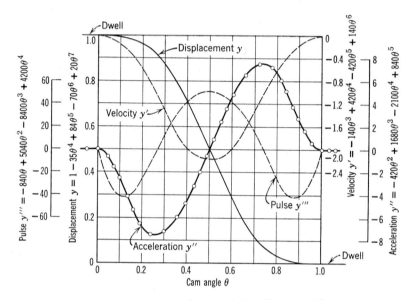

Fig. 6.9. The 4-5-6-7 polynomial (dwell-rise-dwell cam).

values from the ends of the stroke to the midpoint. Thus we realize that under the present fabrication accuracy any additional controls beyond the 3-4-5 polynomial, even where they appear necessary to a design, will most likely produce a reduction in the dynamic values.

6.12 DIRECT SOLUTION OF HIGHER-ORDER POLYNOMIALS

As the number of terms in the equations increases, solutions of the simultaneous equations become more difficult with greater possibility of error. We shall now show a direct method for eliminating the terms equal to zero and finding the values of the coefficients. First, we do not need the general form of eq. 6.11 if we eliminate the terms equal to zero. We know that when $\theta = 0$, and

$$\frac{d^n y}{d\theta^n} = 0$$

the coefficient for that derivative

$$C_n = 0$$

If we refer to the previous articles, we can observe the terms that have zero coefficients. Table 6.2 lists these terms. Next, we can determine

Table 6.2 Zero Coefficient Terms

Initial Powers	Coefficients and Derivative = 0
0 and 1	—
0 to 3	y'
0 to 5	y', y''
0 to 7	y', y'', y'''

the coefficients of the equations by using Table 6.3. So far the value of the coefficient C_0 has always equalled 1, which is the assumed total lift since, at angle $\theta = 0$, $y = 1$. Thus, Tables 6.2 and 6.3 may be used in finding the equations having boundary condition control at the end points equal to zero.

Table 6.3 Coefficient Evaluations

Polynomial $\quad y = C_0 + C_p\theta^p + C_q\theta^q + C_r\theta^r + C_s\theta^s + \cdots$

$$C_p = -\frac{C_0 qrs \cdots}{(q-p)(r-p)(s-p) \cdots}$$

$$C_q = -\frac{C_0 prs \cdots}{(p-q)(r-q)(s-q) \cdots}$$

$$C_r = -\frac{C_0 pqs \cdots}{(p-r)(q-r)(s-r) \cdots}$$

$$C_s = -\frac{C_0 pqr \cdots}{(p-s)(q-s)(r-s) \cdots}$$

Example

Find the coefficients of the seventh-order polynomial by the methods of this article.

Solution

The polynomial is

$$y = C_0 + C_1\theta + C_2\theta^2 + C_3\theta^3 + C_4\theta^4 + C_5\theta^5 + C_6\theta^6 + C_7\theta^7 \quad (6.29)$$

From Table 6.2, we see that the following terms equal zero from the boundary conditions:

$$\theta = 0, \quad y' = 0, \quad C_1 = 0 \quad \text{and} \quad y'' = 0, \quad C_2 = 0, \quad y''' = 0, \quad C_3 = 0$$

giving

$$y = 1 + C_4\theta^4 + C_5\theta^5 + C_6\theta^6 + C_7\theta^7$$

Substituting in Table 6.3 yields coefficients

$$C_p = -\frac{(1)(5)(6)(7)}{(1)(2)(3)} = -35$$

$$C_q = -\frac{(1)(4)(6)(7)}{(-1)(1)(2)} = 84$$

$$C_r = -\frac{(1)(4)(5)(7)}{(-2)(-1)(1)} = -70$$

$$C_s = -\frac{(1)(4)(5)(6)}{(-3)(-2)(-1)} = 20$$

Thus, the equation is

$$y = 1 - 35\theta^4 + 84\theta^5 - 70\theta^6 + 20\theta^7$$

6.13 POLYNOMIAL SOLUTION FOR DWELL-RISE-RETURN-DWELL CAM

On a simplified basis the polynomial contours for dwell-rise-return-dwell cams could consist of two dwell-rise-dwell shapes. However, as shown in Art. 6.6, a discontinuity or sharp "dip" in the acceleration curve is formed at the point of maximum rise. At low speeds, this dip is not prohibitive, but, at high speeds, wear and vibration results. Therefore, in dwell-rise-return-dwell cams, a finite acceleration and zero pulse at the maximum rise point are suggested. This imposes another boundary condition on the curve.

$$\theta = 0, \quad y''' = 0 \quad \text{with } y'' = \text{a finite value}$$

In this discussion only the dwell-rise portion of the curve is considered. This is because the solution of the other part is similar regardless of a symmetrical or unsymmetrical rise-fall curve. Let us present a typical example using these boundary conditions with the methods for solution similar to the dwell-rise-dwell action.

Example

Find the equation of the simplest polynomial for a dwell-rise-return-dwell high-speed cam having zero acceleration at the initial rise point and zero pulse at maximum rise point.

Solution

The six boundary conditions are

$$\theta = 0, \quad y = 1, \quad y' = 0, \quad y''' = 0$$
$$\theta = 1, \quad y = 0, \quad y' = 0, \quad y'' = 0$$

The basic polynomial fulfilling these conditions is

$$y = C_0 + C_1\theta + C_2\theta^2 + C_3\theta^3 + C_4\theta^4 + C_5\theta^5$$

Using Table 6.2, the equation reduces to

$$y = 1 + C_2\theta^2 + C_4\theta^4 + C_5\theta^5$$

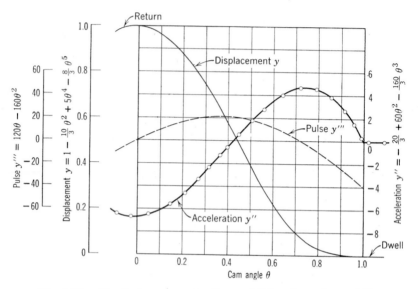

Fig. 6.10. Dwell-rise-return-dwell cam (fifth-order polynomial).

With Table 6.3, the equation becomes

$$y = 1 - \tfrac{10}{3}\theta^2 + 5\theta^4 - \tfrac{8}{3}\theta^5$$

$$y' = -\tfrac{20}{3}\theta + 20\theta^3 - \tfrac{40}{3}\theta^4$$

$$y'' = -\tfrac{20}{3} + 60\theta^2 - \tfrac{160}{3}\theta^3$$

$$y''' = 120\theta - 160\theta^2$$

Plotting, we see Fig. 6.10.

6.14 FITTING THE POLYNOMIAL EQUATIONS TO A GIVEN CURVE

Now we shall show a method of curve fitting for either the displacement, velocity, or acceleration curve by use of polynomial equations. Usually the displacement curve shape is of significance. An example is the automobile cam-valve system, which requires quick opening and closing for best thermodynamic efficiency of the engine. In other cams, the points of maximum velocity or acceleration may be shifted. Furthermore, the maximum positive to negative acceleration ratio may be modified. Additional changes are also possible.

No simplified, direct, exact method is available for the solution of this curve-adjusting process. It is suggested to draw the curves of the various polynomials by arbitrarily varying the powers and coefficients of the polynomials, holding the end conditions equal to zero, and using Tables 6.2 and 6.3. The suggested procedure is:

Start with the basic polynomial, and vary the powers and coefficients until the curve is closely fitted. For example, the fifth-order polynomial could have the powers 3-4-5, 3-7-11, 3-11-18, 3-12-24, etc. The seventh-order polynomial could have powers 4-5-6-7, 4-8-12-15, etc., and the ninth-order polynomial could have powers 5-6-7-8-9, 5-8-13-15-17, 5-9-11-13-15, etc. These calculations may be simplified by use of the Power Tables in Appendix B together with a tabulated work sheet. Other tables are available.[8]

All curves of displacement, velocity, and acceleration should be drawn as a check on the calculation accuracy. Note that extreme difficulty in fitting the curve may indicate poor dynamic properties of the given curve. A new given curve should be applied. Automatic calculators may be utilized to simplify the work.

Example

Take the basic 3-4-5 family dwell-rise-dwell polynomial. Vary the powers to give a larger positive acceleration than the negative acceleration.

Solution

The 3-4-5 polynomial equation (Fig. 6.8) has an acceleration

$$y'' = -60\theta + 180\theta^2 - 120\theta^3$$

Let us try the 3-5-7 and the 3-6-9 polynomials, which by the methods of Art. 6.12 have respective accelerations

$$y'' = -\tfrac{105}{4}\theta + 105\theta^3 - \tfrac{315}{4}\theta^5$$
$$y'' = -18\theta + 90\theta^4 - 72\theta^7$$

Plotting in Fig. 6.11, we see the effect of varying the powers.

6.15 SUBSTITUTION OF ACTUAL CAM EVENTS IN POLYNOMIAL EQUATIONS

Previously we have shown the polynomials with a unit maximum rise in a unit cam angle. Now we shall substitute the actual rise and cam angle to give the particular equations.

If we recall that β = total cam angle from initial point to maximum rise, degrees, and h = total rise in cam angle β, in., it can be shown that conversion to actual conditions is accomplished by multiplying each term of the polynomial $C_n\theta^n$ by the ratio h/β^n, giving

$$hC_n \left(\frac{\theta}{\beta}\right)^n \qquad (6.30)$$

In the preceding paragraphs, we solved the polynomial contours with the boundary conditions at end points either equal to zero or with no

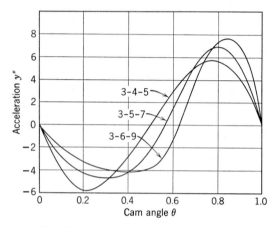

Fig. 6.11.   Polynomial equations of varying powers.

control. It is possible by further study of these equations to give definite values to any of the points of the characteristic curves. For more information it is suggested that the reader again refer to the excellent article on polydyne cam design.[4]

Example

Find the particular displacement, velocity, and acceleration equations for a follower that rises $\frac{1}{2}$ in. in 50 degrees of cam rotation, using the 3-4-5 polynomial.

Solution

In Art. 6.10, the displacement of the basic 3-4-5 polynomial

$$y = 1 - 10\theta^3 + 15\theta^4 - 6\theta^5$$

Multiplying each term by eq. 6.30, $\dfrac{\frac{1}{2}}{50^n}$ yields

$$y = \frac{1}{2} - 5\left(\frac{\theta}{50}\right)^3 + \frac{15}{2}\left(\frac{\theta}{50}\right)^4 - 3\left(\frac{\theta}{50}\right)^5$$

Differentiating gives

$$y' = -\frac{3}{10}\left(\frac{\theta}{50}\right)^2 + \frac{3}{5}\left(\frac{\theta}{50}\right)^3 - \frac{3}{10}\left(\frac{\theta}{50}\right)^4$$

$$y'' = -\frac{3}{250}\left(\frac{\theta}{50}\right) + \frac{9}{250}\left(\frac{\theta}{50}\right)^2 + \frac{3}{125}\left(\frac{\theta}{50}\right)^3$$

Method of Finite Differences

6.16 DETERMINING CHARACTERISTICS

Now we shall consider a numerical procedure for establishing the follower characteristic curves by utilizing incremental tabulated values. These values may be estimated, taken from the actual cam shape measurements, or calculated from an analytical cam curve. This method of finite differences,[9] excellently applied by Johnson,[10] will be employed to:

(a) Determine the velocity and acceleration curves.

(b) Modify the follower acceleration curve to change its shape, improve its dynamic characteristics, and determine the respective displacement values.

The method of finite differences affords us a simple and convenient means for determining the peripheral cutting stages of the actual cam. Furthermore, it permits us for the first time to evaluate the effects of surface imperfections which manifest themselves as local accelerations. The primary shortcomings are that (1) it is an approximation in that the errors increase as the increments are made larger and as the slope of the acceleration curve increases and (2) we cannot find the specific velocity and acceleration values at the dwell end intersecting points of the curves where they are critical. Let

y = displacement of follower, in.
$\Delta\theta$ = incremental cam angle, radians.
Δt = time for cam to rotate through angle, $\Delta\theta$, sec.
v = velocity of follower, ips.
a = acceleration of follower, in./sec^2.

In Fig. 6.12 we see the displacement diagram of a cam having three given points a, b, and c, spaced at small, equal time increments, Δt. The follower velocity of point b

$$v_b = \left(\frac{dy}{dt}\right)_b \cong \frac{y_c - y_a}{2\Delta t}$$

However, we know that $\Delta t = \Delta\theta/\omega$ which, upon substitution, yields:

$$v_b \cong \frac{\omega}{2\Delta\theta} (y_c - y_a) \qquad (6.31)$$

Let us establish the acceleration at point b. Consider that points d and e are midway between points a and b, and points b and c,

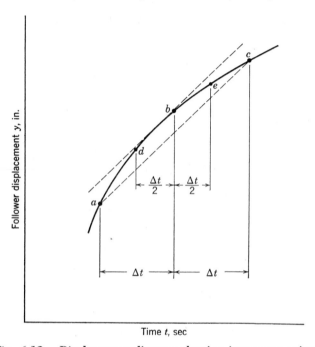

Fig. 6.12.   **Displacement diagram showing increment points.**

respectively. The follower acceleration at point b

$$a_b = \left(\frac{d^2y}{dt^2}\right)_b \cong \frac{v_e - v_d}{\Delta t}$$

Substituting the displacements y_a, y_b, and y_c in the same manner as previously:

$$a_b \cong \frac{\dfrac{y_c - y_b}{\Delta t} - \dfrac{y_b - y_a}{\Delta t}}{\Delta t}$$

Again substituting $\Delta t = \Delta\theta/\omega$ and solving give the follower acceleration

$$a_b \cong \left(\frac{\omega}{\Delta\theta}\right)^2 (y_a + y_c - 2y_b) \qquad (6.32)$$

Let us denote the term $(y_a + y_c - 2y_b)$ as the acceleration factor, which is proportional to the follower acceleration at point b. Similarly $(y_c - y_a)$ may be called the velocity factor. Equations 6.31 and 6.32 will be employed to determine the follower velocity and acceleration at the midpoint between any two points located at small, equal cam intervals, $\Delta\theta$. The smaller these intervals are, and the higher the accuracy of the displacement values, the more reliable this method of solution by finite differences will be.

Example

In Table 6.4 is given the measured cam profile, y_b, for equal cam increments of 2 degrees. The installed cam rotates at 200 rpm. The total rise of the follower is 0.4000 in. in 30 degrees of cam rotation. Determine the follower velocity and acceleration at each point by tabulating values as shown.

Solution

Using the 10-degree point as a sample the velocity and acceleration are

$$v_b \cong \frac{\omega}{2\Delta\theta} (y_c - y_a) \tag{6.31}$$

$$\cong \frac{200 \times 2\pi}{60 \times 2 \times 2 \times \pi/180} (1244 - 512)10^{-4} = 22.0 \text{ ips}$$

$$a_b \cong \left(\frac{\omega}{\Delta\theta}\right)^2 (y_a + y_c - 2y_b) \tag{6.32}$$

$$\cong \left(\frac{200 \times 2\pi}{60 \times 2 \times \pi/180}\right)^2 [512 + 1244 - 2(880)]10^{-4} = -144 \text{ in./sec}^2$$

We observe that the acceleration fluctuates considerably. Furthermore, the velocity and acceleration values at the 0- and 30-degree points cannot be ascertained by this method because the accuracy of the approximation depends upon the assumption of a continuous curve, a condition which does not exist at these points. The velocities at these two end points may be approximated by plotting and extrapolating the velocity curve. The acceleration may be obtained by a graphical slope differentiation of the velocity curve at these points.

Let us discuss the accuracy of the method in terms of the displacement measurements. If these measurements were given to three significant figures in lieu of the four shown, we may have a considerable change in the calculated accelerations. For example, taking the 14-degree point which is one most greatly affected we would

Table 6.4 Cam at 200 rpm. Rise of 0.4000 in. in 30 Degrees

Cam Angle θ, degrees	Displacement y_b, in. $\times 10^{-4}$	y_a, in. $\times 10^{-4}$	y_c, in. $\times 10^{-4}$	v_b Eq. 6.31, ips	Acceleration Factor $(y_a + y_c - 2y_b)$ $\times 10^{-4}$	a_b Eq. 6.32, in./sec^2
0	0	0	80	—	—	—
2	80	0	144	4.3	−16	−576
4	144	80	320	7.2	+112	+4040
6	320	144	512	11.0	+16	+576
8	512	320	880	16.8	+176	+6320
10	880	512	1244	22.0	−4	−144
12	1244	880	1696	24.4	+88	+3160
14	1696	1244	2124	26.4	−24	−864
16	2124	1696	2504	24.2	−44	−1580
18	2504	2124	2896	23.1	+12	+432
20	2896	2504	3292	23.6	+4	+144
22	3292	2896	3600	21.1	−88	−3160
24	3600	3292	3800	15.2	−27	−973
26	3800	3600	3908	9.2	−23	−830
28	3908	3800	4000	6.0	−4	−144
30	4000	3908	—	—	—	—

have displacement values $y_a = 0.124$, $y_b = 0.170$, and $y_c = 0.212$, giving an acceleration equal to $a_b \cong -288$ in./sec^2. This shows acceleration which is one-third of the original, $a_b \cong -864$ in./sec^2. Thus we see that the calculation of acceleration is especially sensitive to small changes of y values since the quantities $(y_a + y_c)$ and $2y_b$ are apt to be close to each other in magnitude.

6.17 ELIMINATING LOCAL ACCELERATION FLUCTUATION

In the foregoing article we observed that employing the method of finite differences presents (in addition to the acceleration curve shape) the acceleration effect of local surface irregularities. If a cam is worn, poorly designed, or poorly fabricated, distinct local fluctuations in the acceleration curve may be evident. These imperfections could be reduced by manipulating the displacement values, utilizing eq. 6.32. In this equation we shall adjust the acceleration factor $(y_a + y_c - 2y_b)$ which gives an indication of the acceleration curve shape. Adjustment of the initial displacement y_b will be offered to obtain a smooth acceleration curve and eliminate local fluctuations. The suggested practice is to adjust the displacement of the point having the greatest deviation or "bump." From eq. 6.32 we see that modifying the displacement of any point y_b results in changes of the acceleration factor for the three points. Modifying the displacement of point b changes the acceleration factor of this point by an addition of -2 times the increment of displacement. Also, adjacent points a and c have their acceleration factor changed by 1 times the displacement adjustment of point b. Caution is urged in observing the signs of the adjustments.

Example

The cam of the example in the previous article has local accelerations which are again shown in Table 6.5. It is desired to improve

Table 6.5   Cam at 200 rpm.   Rise of 0.4000 in. in 30 Degrees

Cam Angle θ, degrees	Displacement y_b, in.$\times10^{-4}$	Displacement Adjustment, in.$\times10^{-4}$	Acceleration Factor $(y_a+y_c-2y_b)$ $\times10^{-4}$, in.	Sample Calculations
0	0			
2	80		-16	
4	144		$+1\not{1}2+56$	$+112-56=+56$
6	320	-56	$+1\not{6}+128$	$+16-2(-56)=+128$
8	512		$+1\not{7}6+1\not{2}0+84$	$+176-56=+120-36=+84$
10	880	-36	$-\not{4}+68$	$-4-2(-36)=+68$
12	1244		$+\not{8}8+52$	$+88-36=+52$
14	1696		-24	
16	2124		-44	
18	2504		$+12$	
20	2896		$+4$	
22	3292		-88	
24	3600		-27	
26	3800		-23	
28	3908		-4	
30	4000			

the cam contour by removing metal from the cam at significant points in the profile. Indicate the amount of machining necessary.

Solution

The solution to this adjustment is shown in Table 6.5. We see that removing 0.0056 in. from the 6-degree point and 0.0036 in. from the 10-degree point would smooth the acceleration curve somewhat. The cam should be further blended on the scalloped edges of the periphery. Caution in machining is advised.

6.18   CURVE ADJUSTING

In a manner similar to the procedure indicated in the previous article we shall change the shape of any curve. We may consider the operation as one of "curve fitting." Generally the acceleration curve shape is the one requiring modification. We may start with any given displacement curve which is either measured in increments or calculated from an exact mathematical relationship. Next we shall calculate the acceleration factors for each point and eliminate the local acceleration fluctuations if necessary. Then we shall adjust the displacements y_b, one at a time, modifying the acceleration factors accordingly. By varying the acceleration factor the acceleration curve may be made flatter, smoother, unsymmetrical, etc., or have any other desired shape. We shall traverse the points in order. Generally several traverses are necessary to obtain the desired characteristics. The number of traverses and the amount of adjustment depend upon the practice and experience of the designer. Note that if the calculations are made from a basic curve or polynomial curve it is suggested to choose one having no discontinuities that most closely fits the final shape desired. In this manner inaccuracies inherent in the method of finite differences are minimized.

Example

We are given a cam having a parabolic motion rise of 0.3000 in. in 30 degrees of rotation. The cam rotates at 200 rpm. Take 2-degree intervals, and (using finite differences) modify the acceleration curve to improve its characteristics. Note that this dynamically poor curve is chosen for its simplicity and to indicate the accuracy of the method at the point of discontinuity.

Solution

Substituting in eq. 2.36 the displacement values are first tabulated in Table 6.6 for each increment angle. Next the acceleration factor is calculated, tabulated, and plotted in Fig. 6.13. Note the trapezoidal shape of the initial acceleration factor curve. Smaller angle increments would increase the calculation accuracy and the true rectangular-shaped curve would be more closely met.

To improve the characteristics we start with a displacement adjustment of -2 at the 2-degree point. The modification to the acceleration factors of the succeeding points are determined in an effort to obtain a sine-shaped curve. Since it was found that the acceleration factor modification had $+50$ at the last 30-degree point,

Table 6.6. Cam at 200 rpm. Rise of 0.3000 in. in 30 Degrees

θ, degrees	Initial Displacement y_b, in. $\times 10^{-4}$	Displacement Adjustment, in. $\times 10^{-4}$ 1	2	Initial Acceleration Factor $(y_a+y_c-2y_b)$ $\times 10^{-4}$	Modification to Acceleration Factor $\times 10^{-4}$ 1	2	Sample Calculations
0	0	0			-2		
2	27	-2		$+53$	$+57\ +30$		$+53 -2(-2) = +57$, choose $+30$
4	107	-27		$+53$	$+51\ +105\ +60$		$+53 -2 = +51$, $+51 -2(-27) = +105$, choose $+60$
6	240	-45		$+54$	$+27\ +144\ +70$		
8	427	-74		$+53$	$+8\ +156\ +75$		
10	667	-81		$+53$	$-24\ +144\ +70$		
12	960	-71		$+54$	$-27\ +88\ +50$		
14	1307	-38	-8	$+39$	$-32\ +37\ +20$	$+12$	
16	1693	-17	-20	-39	$-77\ -43\ -20$	$-40\ -24$	
18	2040	$+23$	-32	-54	$-77\ -117\ -50$	$-82\ -42\ -50$	
20	2333	$+67$	-40	-53	$-30\ -164\ -70$	$-110\ -46\ -66$	
22	2573	$+94$	-45	-53	$+14\ -174\ -75$	$-120\ -40\ -72$	
24	2760	$+99$	-45	-54	$+40\ -158\ -70$	$-120\ -30\ -70$	
26	2893	$+80$	-50	-53	$+46\ -130\ -60$	$-105\ -5\ -50$	
28	2973	$+50$	-45	-53	$+45\ -65$	$+25\ -25$	
30	3000	0			$+50$	$+5$	

a second reverse traverse is made starting at this point. If desired, further traverses may be taken to make the two end points equal in value. Practically speaking, owing to machining inaccuracies this is probably of doubtful value. Thus, for any point the actual fol-

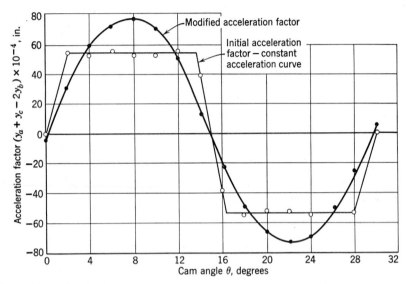

Fig. 6.13. Method of finite differences. Curve adjusting starting with constant acceleration curve. (Cam rotates 200 rpm with rise of 0.3000 in. in 30 degrees.)

lower displacement to be employed in making the cam is the algebraic sum of the initial displacement plus the adjustment. The cam displacement for the 20-degree point is $(2333 + 67 - 32) \times 10^{-4} = 0.2378$ in. Last, it may be mentioned that this method is rapid. Using it, the author required only 30 minutes for the two complete traverses.

6.19 SUMMARY

In this chapter, we have considered the cam curve choice in three ways: (1) combining of basic curves, (2) using polynomial equations, and (3) by the method of finite differences. The essential conditions for combining basic curves are:

(a) The displacement, velocity curves should be continuous functions.

(b) For high-speed action, the acceleration curve should have the lowest maximum value. Its slope (pulse) should be finite and also not too large.

(c) For the rise or fall curves, the area under the positive acceleration portion of the curve must respectively equal the area under the negative acceleration portion of each curve.

Observing these conditions will result in satisfactory performance. We have also shown that the trapezoidal and modified trapezoidal acceleration curve are excellent dwell-rise-dwell curves. For dwell-rise-return-dwell symmetrical curves, the cycloidal curve modified with either the parabolic motion curve or SHM curve is an excellent choice.

The second method, employing polynomial equations, is more direct for both curve fitting and curve shaping. For the dwell-rise-dwell cam, usual practice is to have zero-valued boundary conditions at the initial and final rise points.

The third procedure of finite differences has merit in that it is rapid and provides information on the dynamic effect of cam surface local imperfections.

It should be remembered that with adequate fabrication accuracy all methods can be successfully utilized for cam development to reduce the shock, wear, noise, and vibrations in high-speed, relatively rigid linkages. In the next chapter, we shall design the cam-follower system, compensating for the elasticity of high-speed, highly flexible members.

References

1. A. S. Gutman, "Cam Dynamics," *Mach. Des. 23*, p. 149 (March 1951).
2. D. G. Anderson, "Cam Dynamics," *Prod. Eng. 24*, p. 170 (Oct. 1953).
3. K. G. Holst, "Modified Gravity Curves for Quick Acting Cams," *Amer. Mach. 75*, p. 775 (Nov. 19, 1931).
4. D. A. Stoddart, "Polydyne Cam Design," *Mach. Des. 25*, p. 121 (Jan. 1953); p. 146 (Feb. 1953); p. 149 (March 1953).
5. C. N. Neklutin, "Designing Cams for Controlled Inertia and Vibration," *Mach. Des. 24*, p. 143 (June 1952).
6. C. N. Neklutin, "Vibration Analysis of Cams," *Mach. Des. 26*, p. 190 (Dec. 1954).
7. C. N. Neklutin, "Cams for High Speeds," *Prod. Eng. 5*, p. 250 (July 1934); p. 296 (Aug. 1934).
8. Anon., *Mathematical Tables*, Vol. IX, "Table of Powers Giving Integral Powers of Integers," British Assoc. for the Advancement of Science, Macmillan Co., New York, 1940.
9. M. G. Salvadori and M. L. Baron, *Numerical Methods in Engineering*, Prentice-Hall, Inc., New York, 1952.
10. R. C. Johnson, "Method of Finite Differences for Cam Design," *Mach. Des. 27*, p. 195 (Nov. 1955).
11. C. H. Sexton, "Curve Fitting with Orthogonal Polynomials," *Prod. Eng. 27*, p. 171 (April 1956).

Polydyne Cams

7.1

The polydyne cam combines the polynomial equation with the dynamics of a follower system; the result is an excellent approach to a *high-speed, highly flexible* system. Automobile valve-gear linkages and textile machine members are prime examples. The polydyne method was originally presented by Dudley[1] and elaborated by Stoddart.[2] This approach recognizes that much faulty operation of high-speed, highly flexible systems can be attributed to the difference between "cam command and follower mass response."

The difference in action between the cam and the follower is basically due to elasticity in the linkages; i.e., the system components act as springs of various stiffnesses. Thus, it cannot be assumed that the end of the follower, e.g., the valve of an automobile valve-gear linkage, has the same displacement as the cam profile. In the polydyne method, for the first time we design the *cam shape to give the desired follower action.* This method recognizes that flexibility cannot be eliminated or reduced. Its basic advantages are:

(*a*) By direct means it can eliminate "jump." See Art. 8.10 for a discussion of this phenomenon.

(*b*) By direct calculation it provides the only means of controlling the exact position of the follower end.

(*c*) It limits vibrations to minimum amplitudes if run at the design speed.

Its primary shortcomings are:

(*a*) The high accuracy is needed to realize the advantage of the mathematically computed curve. Sometimes the cam calculated is impossible to cut; i.e. a "dip" in cam is required (Fig. 8.10).

(*b*) The mathematical work is laborious and time-consuming.

Basic Equations

7.2 FUNDAMENTAL RELATIONSHIPS

In the following articles, a typical example indicating the method of attack establishes the basic equations for the high-speed automobile cam valve system (Fig. 7.1). This will show the general approach for any system.

First, let us determine the flexibility relationship of the linkage. Any cam mechanism (assuming a single degree of freedom) may be divided into the usual dynamic equivalent system of four parts.

1. *Compression spring*—to hold the follower on the cam. In positive-drive cams, this spring obviously does not exist.

2. An *equivalent mass* at the end of the follower.

3. A *spring* representing the combined elasticity of the linkage.

4. A *cam*—this may be considered the same as the follower motion for either:

   (a) Low-speed operation.

   (b) Rigid linkage.

Since damping and friction are small, they will be neglected. This will greatly simplify the mathematical relations. Let

k_s = spring rate of compression spring, lb/in.

k_f = spring rate of follower linkage, lb/in.

$m = \dfrac{w}{12g}$ = equivalent mass at the follower end, lb-sec^2/in.

w = equivalent weight at follower end, lb.

L = external load acting on follower, lb.

S_1 = initial compression spring force with mass m at zero position, lb.

N = cam speed, rpm.

y = actual lift at follower end, in.

y_c = rise of cam in. (This is not the same as y because of the linkage deflection).

θ = cam angle of rotation for follower displacement, y, degrees.

Mass m (Fig. 7.1b) is subjected to an acceleration such that at any instant

$$\sum \text{Forces} = m \frac{d^2y}{dt^2} \tag{7.1}$$

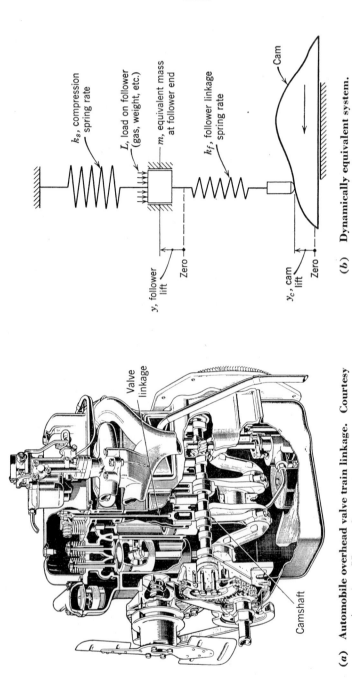

k_s, compression spring rate

L, load on follower (gas, weight, etc.)

m, equivalent mass at follower end

k_f, follower linkage spring rate

Cam

y, follower lift

Zero

y_c, cam lift

Zero

(b)   Dynamically equivalent system.

Valve linkage

Camshaft

(a)   Automobile overhead valve train linkage.   Courtesy American Motors Corp., Detroit, Mich.

Fig. 7.1.   A high-speed, high flexibility cam-follower linkage.

The forces are

$$\text{Main spring force} \ = \ -k_s y$$

$$\text{Linkage force} \ \ \ \ \ \ = \ -k_f (y - y_c)$$

$$\text{External load} \ \ \ \ \ \ = \ -L$$

$$\text{Initial spring force} = \ -S_1$$

Substituting in eq. 7.1 gives

$$-k_s y - L - S_1 - k_f (y - y_c) = m \frac{d^2 y}{dt^2}$$

Solving for cam displacement,

$$y_c = \frac{L + S_1}{k_f} + \frac{k_f + k_s}{k_f} y + \frac{m}{k_f} \frac{d^2 y}{dt^2} \tag{7.2}$$

In eq. 7.2, it is convenient to change the independent variable from time t to the cam angle θ, degrees. We know that

$$\frac{d^2 y}{dt^2} = \omega^2 \frac{d^2 y}{d\theta^2} \tag{2.5}'$$

Substituting gives

$$\frac{d^2 y}{dt^2} = (360)^2 \left(\frac{\text{deg}}{\text{rev}}\right)^2 \times \left(\frac{}{60}\right)^2 \left(\frac{\text{rev}}{\text{sec}}\right)^2 \times \frac{d^2 y}{d\theta^2}$$

$$= 36 N^2 \frac{d^2 y}{d\theta^2}$$

$$= 36 N^2 y'' \tag{7.3}$$

Substituting in eq. 7.2 yields the cam displacement

$$y_c = \frac{L + S_1}{k_f} + \frac{k_f + k_s}{k_f} y + \frac{m}{k_f} \times 36 N^2 y'' \tag{7.4}$$

Rewriting,

$$\underline{y_c = r_a + k_r y + c y''} \tag{7.5}$$

where $r_a = r_s + r_k =$ the ramp height, in. [This is the initial deflection of the follower linkage to eliminate (1) preload, and (2) clearance, so that motion of the follower end is impending].

$$r_s = \frac{L + S_1}{k_f} = \text{initial static deflection of linkage, in.}$$

$r_k = $ clearance or backlash in linkage, in.

$$k_r = \frac{k_f + k_s}{k_f} = \text{equivalent spring rate ratio of the follower linkage.}$$

$$c = \frac{m \times 36N^2}{k_f} = \text{dynamic constant, degrees.}[2]$$

Equation 7.5 may be applied in either of two ways:

(a) Having an arbitrary cam profile y_c, the actual motion of the follower end y may be found. This method, shown in Chapter 8, may be used to investigate existing cam mechanisms.

(b) A suitable follower motion y versus cam angle θ may be assumed, and the cam profile y_c may be developed to fulfill that motion at the desired speed. This is the direct approach used in this chapter.

7.3 RAMP HEIGHT—r_a

The ramp is a small pre-cam and is of critical importance. Its function in height is to compensate for the deflection of the follower due to the clearance r_k and the static deflection r_s. It is the amount that the cam will deflect the linkage before the follower end moves, opposing the preload in the compression spring holding the follower on the cam. In the high-speed, highly flexible systems of this chapter, the clearance r_k is small (held to a minimum) as compared with the static deflection r_s. The ramp height may be found by measurement of the actual linkage or by calculation in which deflection formulae and clearances are applied.

In addition to the clearance, we observe that the ramp height is dependent upon three factors: the external load L, the initial spring force S_1, and the linkage rigidity k_f. Since the external load is often constant and the rigidity of the system is made as large as is permissible, the ramp size is primarily a function of the initial spring load S_1. This in turn is dependent upon the inertia forces as well as other factors.

7.4 SPRING RATIO CONSTANT—k_r

The second term of eq. 7.5 is the constant k_r, called the equivalent spring rate ratio, which is related to the stiffnesses of the follower linkage k_f, and the load spring k_s.

Every part in the linkage acts as a spring of different stiffness. Therefore the follower linkage spring rate k_f in a machine may consist of the sum of many individual springs such as: (a) the bending and twisting of shafts; (b) the deflection and bowing of rods and arms; (c) the deflection of gears; (d) the deflection of bearings; (e) the deflection of cam and follower surfaces.

The most convenient and accurate method for determining the overall value of k_f is by measurement on the actual machine or model. This is done by loading the sys-
tem and accurately observing the deflection with a micrometer indicator. These values when plotted approximate a straight line. The spring rate k_f equals the slope of the plotted line (Fig. 7.2). Another approach for determining k_f is by calculation of above items (a) through (e). This method is subject to error since broad assumptions must be made in the deflection formulae.

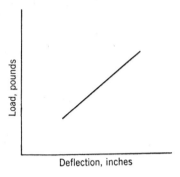

Fig. 7.2.   Spring rate, pounds/inch.

The compression spring rigidity k_s may easily be found by the design calculations, if fabrication accuracy is maintained. In the automotive and aircraft fields, the follower has

$$k_f = \text{between } 20{,}000 \text{ and } 60{,}000 \text{ lb/in.}$$

$$k_s = \text{between } 100 \text{ to } 600 \text{ lb/in.}$$

These values depend on engine size and speed.

7.5 DYNAMIC CONSTANT—c

The dynamic constant

$$c = 36 \frac{m}{k_f} N^2$$

which can be rewritten as

$$c = 0.093 \frac{w}{k_f} N^2 \tag{7.6}$$

where w = equivalent weight at follower end, lb. (Next, let us con-sider the effect of lever arms and gears in a system on the equivalent weight.)

7.6 EQUIVALENT WEIGHT—w

This weight w may or may not be the total weight of the members, depending upon (1) the effect of the extremely flexible members, (2) levers, and (3) gears. The following are suggestions for finding the equivalent weight w:

Springs. For both long thin rods that act as springs and actual compression springs, we do not use their weights directly. Theory and experience have shown that about one-third of their actual weight is effective. This is because flexibility prevents all the mass from being accelerated at the same rate. Thus, the acceleration wave does not affect the other two-thirds of the weight.

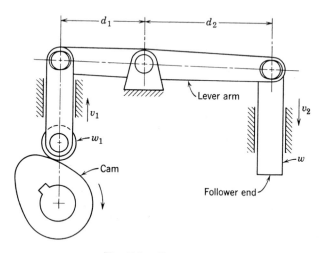

Fig. 7.3.   Lever ratio.

Levers and Gears. A lever (Fig. 7.3) with unequal arms has its effective weight w inversely related to the square of its lever arm ratio. In this figure, we have a lever with arms d_1 and d_2, follower velocities v_1 and v_2, and a weight w_1 to be referred to the follower end.

The kinetic energy of weight w_1

$$K.E. = \frac{1}{2}\frac{w_1}{g}v_1^2$$

This must equal the kinetic energy of the equivalent weight on the follower end

$$K.E. = \frac{1}{2}\frac{w}{g}v_2^2$$

Equating gives

$$w = w_1\left(\frac{v_1}{v_2}\right)^2$$

$$= w_1 U^2 \tag{7.7}$$

where $U = d_1/d_2$ = lever arm ratio. Similarly it can be shown that the equivalent weight or equivalent moment of inertia for rotating gears have the same relationship in which U equals the gear ratio. Therefore any weight or moment of inertia may be referred to the follower end by the square of the lever arm or gear ratio. It may be noted that the inertia of the oscillating arm, being special, may be found by reference to any book on mechanics.

Application

7.7 DISCUSSION

We recall that the cam profile displacement

$$y_c = r_a + k_r y + cy'' \tag{7.5}'$$

Differentiating with respect to θ, yields for the cam,

$$\text{Velocity} \quad = y_c' = k_r y' + cy''' \tag{7.8}$$

$$\text{Acceleration} = y_c'' = k_r y'' + cy^{IV} \tag{7.9}$$

The first four derivatives of the follower motion equation, $y = f(\theta)$, must be continuous functions. This is required, since we desire to maintain continuity of the cam profile y_c, velocity y_c', and acceleration y_c''. Equation 7.9 shows that the cam acceleration y_c'', is a function of the fourth derivative of the follower motion, y^{IV}. Thus, combinations of basic curves cannot be used, since they are discontinuous in these higher derivatives. However, polynomial equations are feasible. In Chapter 6, we have seen that these equations may be utilized to fulfill any boundary condition required simply by adding power parts to the fundamental equation.

$$y = C_0 + C_1\theta + C_2\theta^2 + C_3\theta^3 + \cdots C_n\theta^n \tag{6.11}'$$

The number of terms is dependent on the number of boundary conditions at the end points.

With dwell-rise-dwell action, it can be shown that the 3-4-5 polynomial family for the follower end motion, y, will give finite values of velocity y_c' at the ends. The 4-5-6-7 polynomial indicates finite acceleration values y_c'' at the ends (infinite pulse); vibrations result. This condition can be alleviated by the use of special ramps. However, if we desire that the cam acceleration curve have finite values of pulse, a fourth-order boundary condition [for follower $y = f(\theta)$] equal to zero at the ends is required. From Art. 6.12, the 5-6-7-8-9 polynomial family having the following basic equation fulfills this condition:

$$y = 1 - 126\theta^5 + 420\theta^6 - 540\theta^7 + 315\theta^8 - 70\theta^9 \tag{7.10}$$

This family is suggested for most high-speed, high-flexibility machinery, if the time for calculation and manufacturing accuracy is justified.

The best approach is to establish basic equations with certain simplifying approximations. The significance of the approximations must be compared with the accuracy of the given data and the cutting of the cam. To ignore the effect of flexibility of linkage and the methods of this chapter may result in significant error in performance, analysis, and control of the system. The procedure for design is as follows:

(a) Choose a polynomial equation, $y = f(\theta)$, with proper control at the end points. The 5-6-7-8-9 family is often suggested.

(b) Establish the follower system flexibility relationship, using equations similar to eq. 7.5.

(c) Combine steps (a) and (b), plot displacement, velocity, and acceleration curves of both the cam and the follower end to check the reasonableness of assumptions. This will give the cam shape to be cut. The reader should remember that a comparison between y and y_c must be made after the lever ratio and gear ratio are considered, if they are part of the follower linkage system.

7.8 EXAMPLE USING THE 3-4-5 POLYNOMIAL END MASS MOVEMENT

For simplicity we shall solve a polydyne example, using the impractical 3-4-5 polynomial (for the end mass movement) to indicate the method. We know that the cam will have finite values at the end points, requiring a special ramp to meet these values smoothly. Note that for reasonable accuracy a larger number of significant figures is generally necessary than that shown in the calculations of this problem.

Example

A cam for a high-speed textile machine rotates at 1000 rpm with the follower rising $\frac{3}{4}$ in. in 60 degrees of cam rotation. The follower linkage spring rate is 25,000 lb/in., and the helical compression spring rate is 400 lb/in. The effective weight of the follower (including the lever) is $5\frac{1}{2}$ lb on the cam and $3\frac{1}{2}$ lb on the follower end. The lever arms are 4 in. and 6 in. long (Fig. 7.4). The clearance or backlash in linkage is negligible. The external load is 100 lb, and the initial spring load is 150 lb. Plot curves of the cam and follower displacement, and acceleration. Describe the ramp necessary.

Solution

The equation for basic 3-4-5 polynomial giving the displacement of the follower end

$$y = 1 - 10\theta^3 + 15\theta^4 - 6\theta^5 \qquad (6.20)'$$

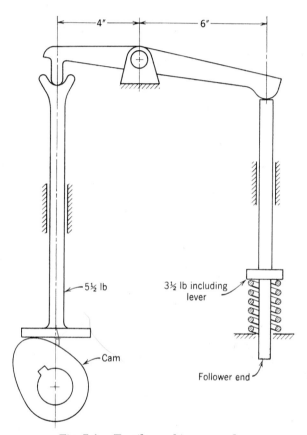

Fig. 7.4.   Textile machine example.

Converting this equation to specific data of this problem according to Art. 6.15 by multiplying by $\dfrac{3/4}{60^n}$ yields

$$y = \frac{3}{4} - \frac{15}{2}\left(\frac{\theta}{60}\right)^3 + \frac{45}{4}\left(\frac{\theta}{60}\right)^4 - \frac{9}{2}\left(\frac{\theta}{60}\right)^5 \qquad (7.11)$$

Differentiating

$$y' = -\frac{3}{8}\left(\frac{\theta}{60}\right)^2 + \frac{3}{4}\left(\frac{\theta}{60}\right)^3 - \frac{3}{8}\left(\frac{\theta}{60}\right)^4 \tag{7.12}$$

$$y'' = -\frac{1}{80}\left(\frac{\theta}{60}\right) + \frac{3}{80}\left(\frac{\theta}{60}\right)^2 - \frac{1}{40}\left(\frac{\theta}{60}\right)^3 \tag{7.13}$$

Now let us find the system characteristics for the cam displacement from eq. 7.5:

$$y_c = r_a + k_r y + c y''$$

The ramp height

$$r_a = r_k + \frac{S_1 + L}{k_f}$$

$$= 0 + \frac{150 + 100}{25,000} = 0.010 \text{ in.}$$

The spring ratio

$$k_r = \frac{k_s + k_f}{k_f}$$

$$= \frac{400 + 25,000}{25,000} = 1.016$$

Assume that the helical spring weighs 0.3 lb. The effective weight

$$w = \left(5\frac{1}{2}\right)\left(\frac{4}{6}\right)^2 + \frac{0.3}{3} + 3\frac{1}{2} = 6.00 \text{ lb}$$

Therefore the dynamic constant from eq. 7.6

$$c = \frac{0.093 w N^2}{k_f}$$

$$= \frac{0.093\,(6.00)\,(1000)^2}{25,000} = 22.2$$

Substituting in eq. 7.5 gives the cam profile

$$y_c = 0.010 + 1.016 y + 22.2 y'' \tag{7.14}$$

Substituting eqs. 7.11 and 7.13 in eq. 7.14 gives the cam displacement

$$y_c = 0.772 - 0.277\left(\frac{\theta}{60}\right) + 0.832\left(\frac{\theta}{60}\right)^2 - 8.175\left(\frac{\theta}{60}\right)^3$$

$$+ 11.43\left(\frac{\theta}{60}\right)^4 - 4.57\left(\frac{\theta}{60}\right)^5$$

Differentiating yields velocity and acceleration

$$y_c{}' = -0.00461 + 0.0277 \left(\frac{\theta}{60}\right) - 0.408 \left(\frac{\theta}{60}\right)^2 + 0.762 \left(\frac{\theta}{60}\right)^3$$
$$- 0.381 \left(\frac{\theta}{60}\right)^4$$

$$y_c{}'' = 0.000461 - 0.0136 \left(\frac{\theta}{60}\right) + 0.0381 \left(\frac{\theta}{60}\right)^2 - 0.0254 \left(\frac{\theta}{60}\right)^3$$

In Fig. 7.5, we shall plot the displacement and acceleration curves of both the cam, y_c and $y_c{}''$, and the follower end, y and y'', respectively. The velocity curves are not shown since they are of lesser

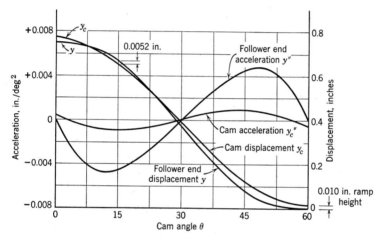

Fig. 7.5. **Textile machine polydyne cam example. Follower end displacement of 3-4-5 polynomial.**

importance. It should be remembered that all these curves are compared on the same basis; i.e., the actual cam would be reduced by the lever arm ratio of 4 in./6 in. or 2/3 scale.

We see that the acceleration curves of the cam $y_c{}''$ has smaller values than the acceleration curve of the follower end y''. This difference is increased as the speed increases. In other words, at a low speed the two curves are practically colinear. In addition, we see an infinite pulse at the ends of the cam acceleration curve $y_c{}''$. This suggests another required function of the ramp for this 3-4-5 curve, i.e., to provide velocity and acceleration values dictated by the cam, in addition to taking up the preload (initial displacement) in the system.

In the displacement curve of the follower, we see the desired 3-4-5 curve shape. The displacement of the cam indicates a ramp of 0.010 in. at the beginning of the curve. However, a serious shortcoming of this design may be observed. At the 15-degree point, we find that the cam displacement, y_c, curve falls below the follower displacement, y, curve about 0.0052 in. If it were possible, the linkage at this point would be in tension. In other words, the cam would leave the follower. This is alleviated by subjecting the linkage to compression at all times. In that event the valve spring force must be increased, a lower speed accepted, or other suitable means utilized. In this problem, we shall use a larger initial spring load S_1 and thus a larger ramp. To recalculate, we have the relationship

$$\frac{S_1 + L}{k_f} = r_s, = r_a \quad \text{since} \quad r_k = 0$$

Here r_a will be increased by 0.006 in. instead of 0.0052 in. for a small margin of safety. Therefore

$$\frac{S_1 + 100}{25,000} = 0.010 + 0.006$$

$S_1 = 300$ lb, the initial spring load with an initial ramp height of 0.016 in. necessary instead of the ramp height 0.010 in. shown in Fig. 7.5.

If we replot this value in Fig. 7.5, the cam curve y_c will properly be above the follower curve y at all points and the linkage would always be in contact with the cam.

7.9   4-5-6-7 AND 5-6-7-8-9 POLYNOMIALS

Taking the data of the previous example and using the 4-5-6-7 and 5-6-7-8-9 polynomials, we have the acceleration curves y'' and y_c'' plotted in Fig. 7.6. With the 4-5-6-7 polynomial, we see that any ramp must meet the cam with an acceleration to prevent an infinite pulse at the beginning and ends of the action. The 5-6-7-8-9 family shows the best condition, having both cam acceleration and follower end acceleration zero at both ends. In contrast, the ramp for this curve must provide only a rise but no initial velocity or acceleration to the follower. This 5-6-7-8-9 family is thus suggested for most designs.

It may be mentioned that higher orders are possible than 5-6-7-8-9 to control the pulse of the follower motion, but they are not justified in view of the additional effort necessary and fabrication inaccuracies that will exist. Furthermore, the changes in contour are negligible, being in the tens of thousandths of an inch.

7.10 RAMPS OR SUB-CAMS

In all polydyne cam designs, it is necessary to bring the linkage to its starting point with a smooth curve. This curve, as previously explained, is needed in many high-speed designs to compensate for the

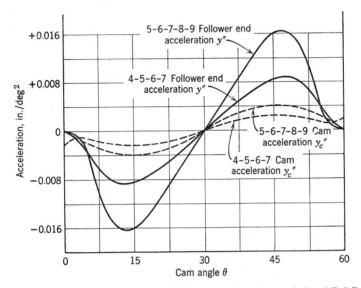

Fig. 7.6. Comparisons of cam and follower accelerations of the 4-5-6-7 and 5-6-7-8-9 polynomials (textile machine example of Art. 7.8).

change in length due to temperature, wear, and flexibility of the parts. Also it is provided in order to start the follower motion, to prevent the follower from leaving the cam, and to provide values of velocity and acceleration to the follower, if necessary. We saw that the 3-4-5 polynomial required a ramp which gave the follower an initial displacement, velocity, and acceleration. We also found that the 4-5-6-7 polynomial needed a ramp for initial displacement and acceleration; the 5-6-7-8-9 polynomial will require only a displacement value from the ramp.

The design of the ramp for polydyne cams is as important as the cam itself. The combination dwell-rise-dwell curves of Chapter 6 could be used excellently to fulfill the 3-4-5 and 4-5-6-7 polynomial ramp requirements. The 5-6-7-8-9 polynomial could be provided with a ramp of the dwell-rise-dwell type, i.e., cycloidal or 3-4-5 type. Some of the ramps employed are: (1) cycloidal; (2) cubic no. 1; (3) combination curves of parabolic motion followed by constant velocity; (4) combination curves of parabolic motion followed by constant velocity and cubic no. 1.

It may be mentioned that in all examples the end points had zero boundary conditions. If finite end point values are needed to suit the ramp or for another reason, it may be fulfilled by further application of the polynomial equations.[2]

7.11 PRACTICAL VIBRATION DISCUSSION

Theoretically, if we run the polydyne cam-follower system at the designed speed, the action will have no vibrations. Actually, small amplitude vibrations (at the natural frequency of the system) are evident in operation. These may be due to the ramp design, the simplified assumptions, the surface inaccuracies, and the application of the external load. The ramp design may not adequately compensate for vibrations. The assumption was made with the equations that damping should be ignored for easier calculations. In actual practice, damping may go as high as 25 percent of critical damping. The surface of the cam profile may have errors in cutting or from wear. Also, the external load may be applied suddenly in machinery such as dial-feed mechanisms. All these factors affect the follower vibratory amplitudes. If a cam is to run at speeds other than rated speed, the best practical approach is to use the harmonic analysis described in Chapter 8, to check resonant conditions. A machine designed by the polydyne method should not be overspeeded since high vibratory amplitudes may be induced for only 10 percent excess of the designed speed.

7.12 SUMMARY

In this chapter, we used an automobile and textile cam-follower system as examples. Other machines could be calculated in the same manner. The methods of this chapter represent a dynamic analysis of a high-speed, high flexibility cam design. Vibrations at the design speed thus approach zero. We also see that any polynomial may be used, but the 5-6-7-8-9 is suggested with a 3-4-5 polynomial or cycloidal ramp. Note that proper ramp design is probably the most critical aspect of the investigation.

Although time-consuming, the justification of this method has been proved in the field. The table of powers in Appendix B and at least a desk calculator are suggested with automatic calculators greatly reducing the work. To repeat the design procedure:

(*a*) Select a polynomial

$$y = C_0 + C_1\theta + C_2\theta^2 + C_3\theta^3 + \cdots C_n\theta^n \qquad (6.11)'$$

(*b*) Set up a cam-follower system whose characteristic is similar to

$$y_c = r_a + k_r y + cy'' \qquad (7.5)'$$

(c) Substitute values, and plot displacement curves and their derivatives.

Be sure that the compression spring is strong enough and the ramp high enough so that cam-follower constraint is maintained. Also, the maximum cam acceleration should always exceed the maximum follower acceleration; otherwise the system may go out of control because it is too flexible.

Finally, a system with a rigid cam shaft and a flexible follower was employed. Sometimes, in cam-driven mechanisms such as turret or dial cams, the cam shaft is light in weight having a small inertia or flywheel effect. In addition, the follower linkage stiffness is high compared with the flexibility of the drive and support frame. Accordingly it is reasonable to consider the cam shaft and frame in the same manner as that shown in deriving the characteristic equations of eq. (7.5). A flexible torsional system will exist. Tests in the field under actual operating conditions with strain gages and an oscilloscope to measure deflection and vibration (under high speeds) will verify the design.

References

1. W. M. Dudley, "New Methods in Valve Cam Design," *Trans. SAE 2*, p. 19 (Jan. 1948).
2. D. A. Stoddart, "Polydyne Cam Design," *Mach. Des. 25*, p. 121 (Jan. 1953); p. 146 (Feb. 1953); p. 149 (March 1953).

The Dynamics of High-Speed Cam Systems

8.1 INTRODUCTION

In the previous chapters, we have presented the cam curves and their displacement, velocity, and acceleration characteristics. Noise, wear, shock, and vibration were mentioned, but in a cursory manner. Now, let us analyze these curves and others at high speeds and study the dynamics of their operation. The higher the speeds, the more critical must be the fabrication accuracy of the contour and the investigation of the proper cam choice. A great improvement in cam operation will result when the importance of surface accuracy and close tolerances is recognized.

In dynamics, the familiar inertia force that is known to be proportional to the acceleration is

$$F_i = ma$$

where F_i = inertia force, lb.

m = mass, lb-sec^2/in.

a = acceleration, in./sec^2.

Many engineers have the misconception that acceleration is the only factor affecting the dynamic reaction under high-speed operation. We shall show that there are other factors. We shall see that shape and continuity of the acceleration curve are of greatest significance in maintaining smooth cam-follower action and that serious vibratory forces occur at points where the acceleration curve has a discontinuity, i.e., infinite pulse.

The conventional cam design is inadequate because it assumes complete rigidity of cam and linkages, without regard to the vibration and deflection of the parts during high-speed action. In fact, owing to the elasticity and the mass of the follower linkage, the follower does not move exactly as the cam shape dictates.

All systems are non-rigid which condition becomes serious if any of the following are encountered: high speeds, low stiffness, high mass, or resonance.

8.2 SOURCES OF VIBRATIONS

In cam-follower systems, vibrations of some sort are always induced. At low speeds, they are rarely a factor for concern, but, at high speeds, they become significant.

Vibrations are of many kinds:

(a) Vibrations due to the *shape* of the follower *acceleration curve.* An infinite slope or pulse ($da/dt = \infty$) will be seen to be undesirable. With the compression-spring-loaded follower, this is called "jump." The transient vibrations that result are more serious in highly flexible linkages.

(b) Vibrations that are a result of *separation* of the *cam* and the *follower.* With positive-drive cams with backlash, impact of the roller on the cam is produced, and is called cross-over shock. With the spring-loaded follower it is due to the "jump" condition.

(c) Vibrations due to *surface irregularities,* of which there are many kinds. This subject is discussed in Chapter 10.

(d) Vibrations due to the *rate* of *application* of the *external load.* For example, a cam-driven punch indexing mechanism has its load applied suddenly as the punch starts into the workpiece. This quick-load application cannot be eliminated, and therefore the ultimate design must include it. Timoshenko[1] has indicated a method of solving this problem. Sometimes the application of load during the minus acceleration period tends to reduce and may even eliminate the reversal of forces acting on the cam surface.

(e) Miscellaneous sources. Vibrations due to *cam unbalance.* Intelligent practical design of the cam structure and the body to reduce the offset mass will keep this vibration to a minimum. Vibration may be transmitted to the cam surface from the *driving mechanism* through the frame from such sources as electrical motors, gears, and chains. Vibrations may be transmitted from an *external source* through the foundation of a machine base or the body structure of an airplane.

To reduce vibrations in the cam-follower system, it is suggested that the members from the driver to the follower end be made as rigid as possible. Damping devices are, as a rule, not suggested as a means of reducing vibrations since damping forces may be comparable to the undesirable inertia forces. In the following articles, we shall elaborate on the foregoing kinds of vibrations, all primarily under high-speed operation.

8.3 VIBRATORY COMPARISON OF BASIC CURVES (DWELL-RISE-DWELL CAM)

Now let us discuss a few of the basic curves in an attempt to establish a proper foundation for their high speed application. Since the parabolic curve gives the smallest value of acceleration and inertia force, it was chosen for years as the best one possible. Theoretically,

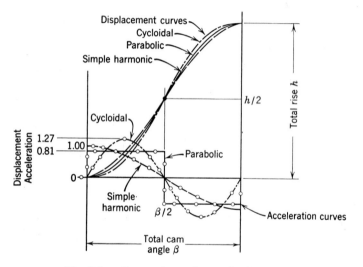

Fig. 8.1.   Comparison of some basic curves.

the parabolic curve would be best with the follower linkage massless or perfectly rigid and without backlash in the system (Fig. 8.1). However, under high-speed operation, this curve is a poor choice. It was found that this curve gave rapid surface destruction, wear, noise, vibrations, and shock at high speeds.

Hrones[3] mathematically analyzed (Art. 8.5), and Mitchell[2] tested, the vibrations resulting from three basic dwell-rise-dwell curves with a high rigidity system (follower). They investigated the transient dynamic magnification factor D, providing the measure of the spring-dashpot coupling effect on the follower. This factor gives the instantaneous ratio of *actual force* F_a to *inertia force* F_i. Stresses and deflections are proportionally related (Fig. 8.2).

The parabolic curve (Fig. 8.2a), with its infinite pulse or discontinuities existing at the ends and midpoint of the stroke, oscillates the follower system at its natural frequency from the first instant of action.

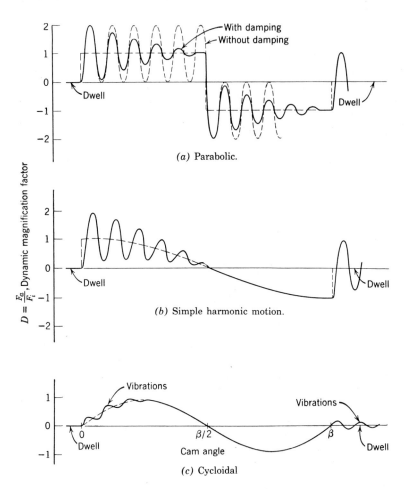

(a) Parabolic.

(b) Simple harmonic motion.

(c) Cycloidal

Fig. 8.2. Vibratory response of basic dwell-rise-dwell curves.

The dynamic magnification factor D is initially equal to 2 at all points of infinite pulse. It will be noted that without damping and with resonance the wave magnitude D theoretically may increase to a value of 3 at the transition point.

With the simple harmonic motion curve (Fig. 8.2b), the oscillations occur at the same ratio of 2 to 1 as the parabolic motion curve, since again we have infinite pulse (discontinuity) at the dwell points.

However, the cycloidal curve (Fig. 8.2c) with its finite pulse at all times has small amplitude vibrations at a maximum ratio of 1.06 to 1. Again the vibrations are at the natural frequency of the system.

Thus the infinite pulse means a sudden or transient application of the inertia load F_i, which produces a shock (twice the value of the inertia load) in the cam-follower system. In design, this phenomenon is often called a "suddenly applied load." We know that these vibrations become more serious as the frequency of this impressed force approaches small, odd-integer multiples of the natural frequency of the system. High-amplitude vibrations and high stresses in resonance and "jump" result. It was shown that the resonant condition is much less affected in the cycloidal than in all other basic curves. We must remember that the fabrication tolerances of the cycloidal cam are critical at the beginning and the end of the stroke and that magnitudes of ±0.0003 in. may be required for good performance.

Sometimes the question is asked, what pulse value should be used for minimum vibratory effects. Neklutin[6] shows by calculations that the modified trapezoidal acceleration curve and the cycloidal curve give excellent characteristics. The vibratory effects are very similar, with the maximum acceleration values of the modified trapezoidal curve slightly smaller than those of the cycloidal curve. By the methods of Art. 8.5, we could show on a theoretical basis that further reduction in pulse at the dwell ends of these curves does not necessarily reduce vibrations. That is, side sloping of the acceleration curve generally has no effect in reducing vibratory amplitudes. Thus blending of the acceleration curve with the dwell ends (zero pulse boundary conditions) does not improve the action.

The conclusions that we can offer for any cam curve at moderate to high-speed operation are:

(a) The acceleration maximum values should be kept as small as possible.

(b) A finite pulse should be maintained at all times, never exceeding the maximum of the cycloidal curve.

(c) The cycloidal curve is a reasonable choice in most cases, giving lower peak forces, vibratory amplitudes, noise, and stresses, and, in general, smoother performance.

8.4 METHODS FOR DETERMINING FOLLOWER RESPONSE

In the next few articles, we shall indicate three methods for establishing the vibratory response of a follower linkage of any rigidity. Note that in all discussions we shall not mention the resonant effect of infinite pulse values related to each other since any infinite pulse at all is undesirable at high speeds. It is suggested that the reader use caution in choosing the method that most closely applies to his problem. All methods have been verified in practice where good engineering judgment has been applied.

In Fig. 8.3, we see that all cams have repeated cyclical action of any rise, fall, and dwell proportions. The reader should refer to this figure for the discussions that follow.

The first method for determining the follower response is *analytical* and may be conveniently applied when the cam profile is given in mathematical form. This approach gives the transient response start-

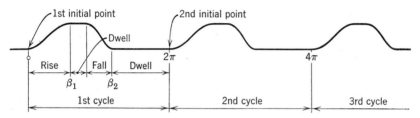

Fig. 8.3.   Cam cycle.

ing at each initial point (the dwell and curve intersection) of each cycle or part of a cycle. It assumes no vibratory effect from any previous cycle. Damping and the dwell period of the previous stage or cycle are predominant factors in eliminating vibrations that occur before the cycle under consideration. This damping effect is improved as the natural frequency of the system is made higher. The basic shortcoming of this analytical method is that the response may be conveniently found only for the period β of the analytical expression, e.g., with a basic or polynomial curve between the dwell periods. Sometimes the action that follows in the dwell range cannot be easily investigated.

The second method is a *numerical* one, using a cam acceleration curve which is given either graphically, numerically, or not in convenient analytical form. As before, this approach gives the transient vibratory response. However, in contrast, it is an improvement in that the complete cam function, including dwells, can be easily investigated. This method is best applied where an adequate dwell period exists at the end of the fall action for each cycle (transient condition). In this manner, there will be no vibrations affecting the next initial starting point.

The third approach is the solution of a steady-state condition employing Fourier series or *harmonic analysis*. We may superimpose the effects of each harmonic or make a comparison between high amplitudes of the lower harmonics and the natural frequency of the system. This method neglects the initial transient condition. It assumes that the free vibrations of the system usually generated at the beginning of motion have been damped out by friction, and forced vibrations alone are considered. Its application is suggested when the cam rise period β_1 and fall period β_2 are a large portion of the total cam cycle (2π) in which the vibratory waves from each cycle, 2π, 4π, 6π, etc., reinforce each other. We have to concern ourselves with the cyclic or repeated effect of the dynamic forces and the vibrations that remain after the stroke because these have an influence on the magnitudes of the vibrations in the next stroke. The Laplace transform may be applied in some cases to provide the transient and steady state solution of the follower response. In all installations, since damping cannot be easily increased, it is suggested that the natural frequency of the linkage be as high as possible to reduce the vibratory period for the same percentage of damping. Last it may be mentioned that the follower response may be determined experimentally by employing high-speed photography[20,21] and oscillograph measurements.[22] The former method is applicable to highly elastic systems.

8.5 ACTUAL MOTION OF FOLLOWER— ANALYTICAL METHOD

Now let us analyze the exact follower motion for any system having the cam profile in convenient analytical form. Figure 8.4 shows the equivalent mass m and the follower at rest when $t = 0$. The relationship[3] assuming a single degree of freedom is

$$\sum \text{Forces} = m\frac{d^2y}{dt^2} \tag{8.1}$$

or

$$-b_l\left(\frac{dy}{dt} - \frac{dy_c}{dt}\right) - b_f\frac{dy}{dt} - k(y - y_c) = m\frac{d^2y}{dt^2} \tag{8.2}$$

This gives

$$\frac{dy^2}{dt^2} + 2(\eta_l + \eta_f)\omega_n\frac{dy}{dt} + \omega_n^2 y = 2\eta_l\omega_n\frac{dy_c}{dt} + \omega_n^2 y_c \tag{8.3}$$

where, $\omega_n = (k/m)^{\frac{1}{2}}$ which is the undamped natural frequency of the linkage, radians/sec. (8.4)

or

$$\left[f_n = \frac{1}{2\pi}(k/m)^{\frac{1}{2}} \quad \begin{array}{l}\text{which is the undamped natural fre-}\\ \text{quency of linkage, cycles/sec.}\end{array}\right] \tag{8.5}$$

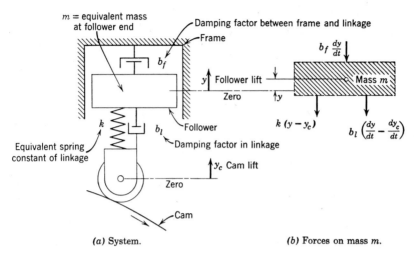

(a) System.   (b) Forces on mass m.

Fig. 8.4.   Equivalent system of cam-follower mechanism.

Note that, for practical purposes, the damped and the undamped natural frequencies are the same.

$$\eta_l = \frac{b_l}{2(km)^{\frac{1}{2}}} = \text{damping ratio in the follower linkage, lb-sec}^2/\text{in.} \quad (8.6)$$

$$\eta_f = \frac{b_f}{2(km)^{\frac{1}{2}}} = \begin{array}{l}\text{damping ratio between frame and linkage,} \\ \text{lb-sec}^2/\text{in.}\end{array} \quad (8.7)$$

where m = equivalent mass of the follower, lb-sec^2/in. (Chapter 7)

k = equivalent spring rate of linkage, lb/in.

b_l and b_f = damping factors in linkage and between linkage and frame, respectively, lb-sec/in.

y = follower displacement, in.

y_c = cam displacement, in.

Equations. 8.2 and 8.3 are the general expressions for any system. All that is necessary to find the follower response is to substitute into the analytical expression y_c for the particular cam and solve for the follower characteristics.

If damping is considered negligible for simplicity, eq. 8.2 becomes

$$-k(y - y_c) = m\frac{d^2y}{dt^2} \quad (8.8)$$

Let us determine the response of the simple harmonic motion D-R-D cam in order to indicate the general method of solution. The displacement

$$y_c = \frac{h}{2}\left(1 - \cos\frac{\pi\theta}{\beta}\right) \tag{2.18}'$$

where h = total rise of follower, in.

β = cam angle rotation for rise h, radians.

θ = cam angle rotation for cam rise y_c, radians.

ω = cam angular velocity, radians/sec.

$n = \dfrac{\omega_n}{\omega} = \dfrac{2\pi f_n}{\omega}$ = frequency ratio = 1 at resonance with fundamental.

Substituting eq. 2.18 in eq. 8.8 yields

$$\frac{d^2y}{dt^2} + \omega_n{}^2 y = \frac{\omega_n{}^2 h}{2}\left(1 - \cos\frac{\pi\theta}{\beta}\right) \tag{8.9}$$

This has a general solution (complementary plus particular)

$$y = A\sin n\theta + B\cos n\theta + \frac{h}{2}\left[1 - \frac{1}{1 - \left(\dfrac{\pi}{n\beta}\right)^2}\cos\frac{\pi\theta}{\beta}\right]$$

A and B are arbitrary constants. For this cam curve the boundary conditions are

$$y = y' = 0 \text{ at } \theta = 0$$

so that

$$A = 0$$

and

$$B = \frac{h}{2}\left[\frac{1}{\left(\dfrac{n\beta}{\pi}\right)^2 - 1}\right]$$

The general equations for the displacement, velocity, and acceleration or the response of mass m, at the linkage end are

$$y = \frac{h}{2}\frac{1}{\left(\dfrac{n\beta}{\pi}\right)^2 - 1}\left(-\cos\frac{\pi\theta}{\beta} + \cos n\theta\right) \tag{8.10}$$

$$v = \frac{dy}{dt} = \frac{h}{2}\frac{1}{\left(\dfrac{n\beta}{\pi}\right)^2 - 1}\left(\frac{\pi\omega}{\beta}\sin\frac{\pi\theta}{\beta} - \omega_n\sin n\theta\right) \tag{8.11}$$

$$a = \frac{d^2y}{dt^2} = \frac{h}{2}\frac{1}{\left(\dfrac{n\beta}{\pi}\right)^2 - 1}\left[\left(\frac{\pi\omega}{\beta}\right)^2\cos\frac{\pi\theta}{\beta} - \omega_n{}^2\cos n\theta\right] \tag{8.12}$$

We know that the principal force acting on a system is the accelerating force. In Fig. 8.5, we have plotted the response characteristic curves of the displacement y, velocity v, and acceleration a, superimposed on the respective SHM cam characteristics curves. These are transient response curves for relatively flexible systems, such as aircraft engine valve-gear and textile machines, where we have small values of n, e.g., 2, 3, and 4. For more rigid systems, the curves and analysis will be similar, having higher frequencies and smaller amplitudes. In Fig.

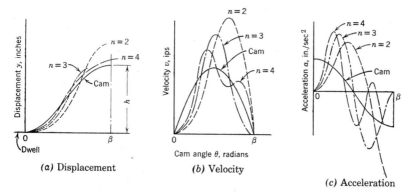

(a) Displacement   (b) Velocity

(c) Acceleration

Fig. 8.5. Response of mass for SHM cam

$$n = \frac{\omega_n}{\omega} \text{ (no damping assumed).}$$

8.5a, the mass catches up and overshoots the forcing cam displacement. In other words, the dynamic compression of the highly flexible follower linkage causes a subsequent release of energy which in turn causes the end of the follower to surge ahead or behind the cam lift. Actually the amplitudes will be reduced by damping. It is suggested that the natural frequency be made as high as possible so that, if resonance (at a given speed range of cam operation) occurs, it will be with a higher frequency ratio and consequently lower vibratory amplitudes. Also, the more rigid the system (the higher its natural frequency ω_n), the closer will be the mass response to the cam contour.

In the same manner as that shown for the simple harmonic curve example, the response characteristics can be found for any cam curve and mechanism.

8.6 ACTUAL MOTION OF FOLLOWER— NUMERICAL METHOD

Often an analytical expression for the disturbing force (cam shape) is not known or is not in a convenient form. The cam shape is given graphically, numerically, or made up of combination curves (see Chapter 6). Timoshenko and Young[8] indicate a method of solution. The actual motion of the follower may be found numerically by replacing the cam acceleration curve with a step curve having a constant time interval Δt. The cam disturbing force can be considered a succession of constant forces ma_1, ma_2, ma_3 $\cdots$ (proportional to static deflections $a_1/\omega_n{}^2$, $a_2/\omega_n{}^2$, $a_3/\omega_n{}^2$ $\cdots$), each acting for this time interval. With the initial displacement $y_0 = 0$ and initial velocity $dy_0/dt = 0$, we can find the follower displacement and velocity at the end of the first interval due to this cam force:

$$y_1 = y_0 \cos \omega_n \, \Delta t + \frac{dy_0/dt}{\omega_n} \sin \omega_n \, \Delta t + \frac{a_1}{\omega_n{}^2} (1 - \cos \omega_n \, \Delta t) \quad (8.13)$$

$$\frac{dy_1/dt}{\omega_n} = -y_0 \sin \omega_n \, \Delta t + \frac{dy_0/dt}{\omega_n} \cos \omega_n \, \Delta t + \frac{a_1}{\omega_n{}^2} \sin \omega_n \, \Delta t \quad (8.14)$$

We shall take these values of y_1, and (dy_1/dt) as new initial displacements and velocity for each step and substitute in the same manner for the second interval. In this way, having values of y_1, y_2, y_3 $\cdots$ corresponding to times 0, Δt, $2\Delta t$, $\cdots$ gives the follower response curve. The total displacement of the end of the follower linkage is made up of:

(1) Cam displacement curve y_c.

(2) Plus or minus vibratory response of the end of the linkage (discussed above).

(3) Minus the backlash or clearance in the system. On positive-drive followers, it is added in the negative acceleration range and subtracted in the positive acceleration range. For simplicity ignore this factor.

(4) Minus the linkage deflection due to the change in compression spring loading. This is calculated by dividing the spring force at every point (see Chapter 9) by the equivalent spring scale of the linkage. This factor is generally negligible.

(5) Plus or minus the miscellaneous deflections due to external loads, impact in operation, etc. The magnitudes are often too complex to be considered.

Damping may be included in these relations if desired. Lastly, it should be remembered that, the smaller the intervals of Δt, the more accurate will be the response curve. The foregoing numerical method has been verified in practice and provides acceptable accuracy in determining the actual follower motion.

Example

A positive-drive cam with no backlash in the system rotates at 300 rpm. A symmetrical trapezoidal acceleration curve is to be utilized. The total rise is 5.4 in. in 160 degrees of cam rotation. A relatively flexible follower linkage with its mass principally located at its end has a natural frequency of 3000 cycles/min. Find the displacement of the follower for the complete action.

Solution

The time for one revolution = $\frac{60}{300}$ = 0.2 sec.

The time for 80 degrees of revolution = $0.1 \times \frac{80}{360}$ = 0.0444 sec.

Article 6.4 suggests that the constant pulse period of the acceleration curve is $\frac{1}{4} \times 80$ = 20 degrees or 0.0111 sec. By the methods of Chapter 6, we find a maximum acceleration of 2050 in./sec². In Fig. 8.6a are plotted the cam acceleration curve and the follower natural frequency cycle and period. The natural frequency of the follower linkage

$$\omega_n = 3000 \times \frac{2\pi}{60} = 314 \text{ radians/sec}$$

Let us take ten divisions of the cam acceleration curve, each increment Δt = 0.0044 sec, being shown plotted in steps. In Table 8.1, we tabulate values, using the following procedure to find the response curve:

(a) Fill out the first five columns completely.

(b) Put y_0 and $(dy_0/dt)/\omega_n$ both equal to zero in the first line of columns 6 and 9.

(c) Using eqs. 8.13 and 8.14, the values of y_1 and $(dy_1/dt)/\omega_n$ are computed and placed in the second line of columns 6 and 9.

(d) The remaining spaces of the second line are filled in.

(e) Similarly, the values of y_2 and $(dy_2/dt)/\omega_n$ can be calculated by eqs. 8.13 and 8.14 of the next interval, and recorded in the third line of columns 6 and 9.

(f) In Fig. 8.6b are plotted the values of columns 3 and 6, giving the static deflection $a/\omega_n{}^2$ (this curve is shown primarily as a check) and the follower response curves. For example at the end of the third interval the follower linkage will be compressed y equal to 0.0150 in.

(g) The actual displacement curve (not shown) may now be plotted or values tabulated.

Note that damping would reduce these amplitudes proportionally for each cycle.

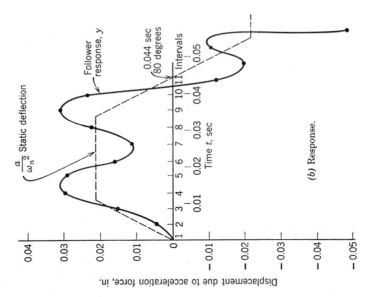

(b) Response.

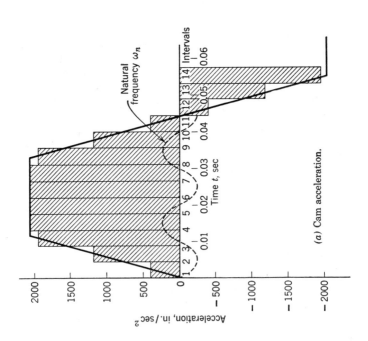

(a) Cam acceleration.

Fig. 8.6. Numerical method for finding follower response—trapezoidal acceleration curve cam.

Table 8.1 Numerical Solution for Follower Response

$\omega_n = 314$ rad/sec, $\omega_n{}^2 = 98{,}800$, $\Delta t = 0.0044$ sec, $\omega_n \Delta t = 79.3$ degrees, $\sin \omega_n \Delta t = 0.983$, $\cos \omega_n \Delta t = 0.186$, $1 - \cos \omega_n \Delta t = 0.814$

Interval	(1) t, sec	(2) a, in./sec^2	(3) $\dfrac{a}{\omega_n{}^2}$, in.	(4) $\dfrac{a}{\omega_n{}^2} \times (1-\cos\omega_n\Delta t)$, in.	(5) $\dfrac{a}{\omega_n{}^2} \times \sin\omega_n\Delta t$, in.	(6) y, in.	(7) $y\cos\omega_n\Delta t$, in.	(8) $y\sin\omega_n\Delta t$, in.	(9) $\dfrac{dy/dt}{\omega_n}$, in.	(10) $\dfrac{dy/dt}{\omega_n}\times\cos\omega_n\Delta t$, in.	(11) $\dfrac{dy/dt}{\omega_n}\times\sin\omega_n\Delta t$, in.
1	0	......	......	....	....	0	0	0	0	0	0
2	0.0044	+ 400	+0.0040	+0.0033	+0.0040	+0.0033	+0.0006	+0.0032	+0.0040	+0.0007	+0.0039
3	0.0088	+1280	+0.0129	+0.0105	+0.0127	+0.0150	+0.0028	+0.0148	+0.0102	+0.0019	+0.0100
4	0.0132	+1950	+0.0198	+0.0161	+0.0194	+0.0289	+0.0054	+0.0284	+0.0065	+0.0012	+0.0064
5	0.0176	+2050	+0.0208	+0.0169	+0.0204	+0.0287	+0.0053	+0.0282	-0.0068	-0.0013	-0.0067
6	0.0220	+2050	+0.0208	+0.0169	+0.0204	+0.0155	+0.0029	+0.0152	-0.0090	-0.0017	-0.0089
7	0.0264	+2050	+0.0208	+0.0169	+0.0204	+0.0109	+0.0020	+0.0107	+0.0035	+0.0006	+0.0035
8	0.0308	+2050	+0.0208	+0.0169	+0.0204	+0.0224	+0.0042	+0.0220	+0.0103	+0.0019	+0.0101
9	0.0352	+1950	+0.0198	+0.0161	+0.0194	+0.0304	+0.0057	+0.0300	-0.0007	-0.0001	-0.0007
10	0.0396	+1280	+0.0129	+0.0105	+0.0127	+0.0156	+0.0029	+0.0152	-0.0173	-0.0032	-0.0170
11	0.0440	+ 400	+0.0040	+0.0033	+0.0040	-0.0108	-0.0020	-0.0106	-0.0144	-0.0027	-0.0142
12	0.0484	- 400	-0.0040	-0.0033	-0.0040	-0.0195	-0.0036	-0.0191	+0.0040	+0.0007	+0.0039
13	0.0528	-1280	-0.0129	-0.0105	-0.0127	-0.0103	-0.0019	-0.0101	-0.0311	-0.0058	-0.0306
14	0.0572	-1950	-0.0198	-0.0161	-0.0194	-0.0486					
15											

8.7 ACTUAL MOTION OF FOLLOWER—
HARMONIC ANALYSIS

In Art. 8.5, we have shown the general expression for a cam-follower system to be

$$m\frac{d^2y}{dt^2} + (b_l + b_f)\frac{dy}{dt} + ky = b_l\frac{dy_c}{dt} + ky_c$$

or

$$m\frac{d^2y}{dt^2} + (b_l + b_f)\frac{dy}{dt} + ky = F(t) \qquad (8.15)$$

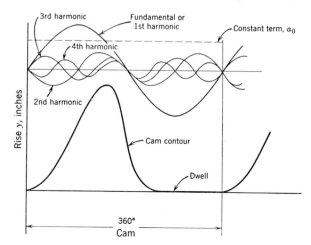

Fig. 8.7.   Harmonic analysis (resolving into sinusoidal components).

showing a periodical disturbing force as a function of time, $F(t)$. This force can be represented in the form of a trigonometric series of sinusoidal functions,[1,9,23] such as

$$m\frac{d^2y}{dt^2} + (b_l + b_f)\frac{dy}{dt} + ky$$

$$= a_0 + a_1\cos\omega t + b_1\sin\omega t + a_2\cos 2\omega t + b_2\sin 2\omega t + \cdots \qquad (8.16)$$

In Fig. 8.7, we see the harmonic analysis of the first four components of a cam contour where the harmonic numbers are 1, 2, 3, and 4. The coefficients a and b can be calculated if $F(t)$ is known analytically. As before, the general solution of this equation will consist of two parts, one free and the other forced vibration. Steady-state solution may be found by ignoring the free vibrations and solving for the follower response by superimposing the effects of the magnitudes of components

$a_1, b_1, a_2, b_2, a_3, b_3, \cdots$ at each point. Also, large forced vibrations may occur when the period (sec) of one of the terms of the series coincides with (or is a multiple of) the period (sec) of the natural frequency of the system. This is called resonance, which occurs at the critical speed of the respective harmonic number, i.e., when n equals an integer.

Furthermore, the magnitude of the response at resonance will depend primarily on the magnitude of the components $a_1, b_1, a_2, b_2, a_3, b_3, \cdots$ for that harmonic. Vibration is caused by a force, and thus the harmonics of the acceleration curve rather than the displacement curve are responsible for it. For any contour, the acceleration and displacement harmonics are proportional and the latter are usually considered.

Harmonic analysis of the acceleration forces produced by a given cam design will give an indication of the performance to be expected. As before, the natural frequency ω_n should be as high as possible so that resonance at a given speed range will occur with a higher harmonic number of consequently smaller amplitude harmonics. Fortunately only a few of the low numbers cause excessive vibratory amplitudes. Some of the higher numbers appear as noise and may be objectionable. Lastly, the amplitude of the forced oscillation is small if the frequency of the external force is different from the natural frequency and, of course, depends on the proximity to the natural frequency of the system.

Also, it should be remembered that a smooth "bumpless" acceleration curve generally gives weaker harmonics and thus a smaller response. By smooth is meant few points of inflection, i.e., the D-R-D acceleration curve has one point of inflection. A refinement would be to have all derivatives of the acceleration curve d^3y/dt^3, $d^4y/dt^4 \cdots$ continuous functions. However, in actual practice it is impossible to fabricate the cam to the accuracy demands of these higher derivatives, and their practical value has never been verified.

If $F(t)$ (cam curve) is given numerically or graphically because no analytical expression is available, some approximate numerical method for calculating harmonics can be obtained by using one of the instruments for analyzing curves in trigonometric series. A discussion of this subject may be found in Jehle and Spiller.[10]

Example*

A positive-drive radial cam turning once every 4 sec operates a roller follower on a push rod. This rod moves a rocking lever and, by means

* Courtesy of F. R. E. Crossley.[7]

of a connecting link, a sliding table weighing 185 lb. The displacement
of the roller follower from dwell to dwell is as follows:

Cam angle, degrees	10	20	30	40	50	60	70	80	90
Displacement, in.	0.080	0.215	0.439	0.730	1.112	1.500	1.898	2.250	2.561

Cam angle, degrees	100	110	120	130	140	150	160	170	180
Displacement, in.	2.789	2.949	3.049	3.155	3.410	3.781	4.132	4.422	4.500

The strength of the connecting members are: under 100 lb of compressive load, the push rod deflects 0.0012 in., the connecting link
0.0165 in., and the rocking lever bends 0.237 in. Masses of these members are small in comparison with the table. While the cam is operating, a troublesome chatter develops in the table. Find the cause.

Solution

The overall (equivalent) spring constant for a system in series

$$\frac{1}{k} = \frac{1}{k_1} + \frac{1}{k_2} + \frac{1}{k_3} + \cdots$$

$$= \frac{1}{100/0.0012} + \frac{1}{100/0.0165} + \frac{1}{100/0.237}$$

$$k = 392 \text{ lb/in.}$$

From eq. 8.5, the natural frequency of the follower

$$f_n = \frac{1}{2\pi} \left(\frac{k}{m} \right)^{1/2} = \frac{1}{2\pi} \left(\frac{392}{185/386} \right)^{1/2} = 4.55 \text{ cycles/sec}$$

Plotting the displacement of the follower in Fig. 8.8, we can by
employing the method of finite differences find the velocity and
acceleration curves. By observation and by harmonic analysis (not
shown), we find a strong imposed frequency of acceleration of 4½
cycles/sec or 18th harmonic number. It may be seen that the frequency ratio

$$n = \frac{f_n}{\omega/2\pi} = \frac{4.55 \text{ cycles/sec}}{\frac{1}{4} \text{ rev/sec}} = 18$$

Thus, by observation of Fig. 8.8, we see that the chatter is caused by
(a) resonance between natural frequency and the strong eighteenth
harmonic number; (b) excessive pulse values, especially at the dwell
ends of the cam curve. A smoother acceleration curve with smaller
maximum pulse values solved the problem by yielding weaker higher
harmonics.

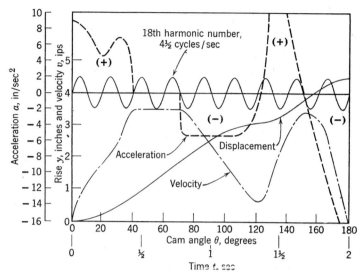

Fig. 8.8. Cam example of Art. 8.7 (chatter resulted from resonance with strong 18th harmonic number).

8.8 PRACTICAL DISCUSSION ON DETERMINING FOLLOWER RESPONSE

In the foregoing articles, we have shown three methods for establishing the response of a cam follower. Now let us discuss some practical factors that may predominate over the mathematical analysis previously presented.

First, the systems were considered on the basis of a single degree of freedom in which the machine frame was perfectly rigid. In general, the frames are stiff enough to limit the deformations to negligible amplitudes. However, in the range of resonance the values may have magnitudes requiring investigation when the cam speed agrees with the natural vibratory mode. Tests of the natural frequencies of the frame may be necessary. More exact methods may require the basis of two or more degrees of freedom—generally, calculating time and effort may be prohibitive.

Also, all surface irregularities will affect the cam in that there is frequently a large difference between the theoretical and the actual cam acceleration curves. This affects the follower acceleration and motion, and cannot be overemphasized. Accuracies as high as ± 0.0005 in. in cam profile are often necessary. See Chapter 10 for further discussion of this subject. Surface irregularities present a significant factor in any cam-follower vibration discussion. An effort was therefore made to produce lift curves and cam contours which would primarily facilitate accurate manufacturing. Schlaefke,[11] on the basis of calculations showing the influence of machining accuracy, suggested a follower motion with an acceleration curve of parabolic shape.

Furthermore, the two significant related factors, cam speed and natural frequency of the cam-follower linkage, may vary during operation. First, the forced vibrations which depend upon the cam speed vary slightly with the prime mover. For example, electrical equipment is affected by voltage variations under industrial conditions. Secondly, the natural frequency of a machine is not exact, being affected by radial clearance in bearings, frictional resistance of metals in contact and lubricant, changes in load handled by the machine, dynamic oil film effect, and the bowing of columns, in addition to the flexibilities and masses of a system. Thus, it may vary slightly during operation.

It should be mentioned that the calculations for natural frequency may be as high as 20 percent in error. A more accurate method uses electrical oscillographic equipment on the actual machine or model.[12] The natural frequency of automobile overhead valve gear linkage is between 40,000 and 50,000 cycles per min and that of aircraft valve gear linkage between 20,000 and 30,000 cycles per min.

Although damping was neglected in the examples of this chapter, refined investigation necessitates its direct inclusion in the equations. We know that damping in most cam-follower assemblies may be between 10 and 25 percent of critical damping.

An attempt to reduce the vibratory amplitudes by increasing the damping action should be used with discretion since the damping forces may become higher and exceed the dynamic force. Barkan[13] shows in his analysis that damping could be included in the calculation for a closer correlation between theory and test. In eq. 8.2, even for a small damping factor $\eta_f = 0.1$, the damping forces are large relative to the inertia forces. In other words, as η_f increases the transient oscillations decrease rapidly but the peak forces increase. For better correlation, damping factors for high-speed engines should be evaluated under actual performance. Although damping devices and their analysis are complicated, they represent a promising field for the study of improving machine performance at high speeds.

8.9 CROSS-OVER SHOCK

In the previous articles, we discussed the vibratory response of the follower mass in a high-speed cam-follower system. Let us now consider a phenomenon called cross-over shock,[7] which exists with positive-drive cams when the contact of the follower shifts from one side to the other. Thus, the clearance or backlash between the follower and the cam are taken up with resulting impact. We know that cross-over shock exists in all positive-drive systems, whether the follower is rigid or flexible. It occurs at a point where the acceleration of the follower changes from positive to negative, or vice-versa. This is also the point of maximum follower velocity. The less this velocity, the less is the shock of the exchange. We see from Fig. 8.5 that the cross-over shock and the maximum velocity (velocity shock) of the mechanism become less as the follower is made stiffer. This cross-over shock can be kept to a minimum by using a rigid follower system or having a high natural frequency. Preloading of the follower by utilizing dual rollers, shown in Art. 5.23, is a practical means of alleviating this detrimental condition—backlash is removed from the system. In addition, it can be shown by mathematical analysis that certain cam curves (such as the cycloidal) having no infinite pulse are best. They provide low cross-over shock because of the inherent small vibratory amplitudes produced. The parabolic curve is poor. In conclusion, this cross-over shock produces additional serious vibrations which are to be superimposed on the curves of Fig. 8.5.

8.10 JUMP PHENOMENON

With the compression-spring-loaded follower, we may have a similar kind of vibration called jump or bounce. This is a transient condition that occurs only with *high-speed, highly flexible* cam-follower systems. With jump, the cam and the follower separate owing to excessively unbalanced forces exceeding the spring force during the period of negative acceleration. This is undesirable since the fundamental function of the cam-follower system, the constraint and control of follower motion, are not maintained. Also related are the short life of the cam flank surface, high noise, vibrations, and poor action.

We can observe from Art. 8.5 that the acceleration response curve is related to the number of full cycles of natural follower vibrations occurring during the initial *positive-acceleration* time interval at the particular operating speed. Defining λ as the number of free vibration cycles per positive acceleration period, we can see that

$$\lambda = \frac{\beta_1}{2\pi} n \qquad (8.17)$$

where β_1 = angle of positive acceleration period, radians (Fig. 8.9). For example, in a high-speed, highly flexible mechanism with the cam-shaft speed of 2100 rpm, the follower natural frequency of 42,000 cycles per minute and $\beta_1 = 27$ degrees, we find the frequency ratio

$$n = \frac{f_n}{\omega/2\pi} = \frac{42{,}000 \text{ cpm}}{2100 \text{ rpm}} = 20$$

From eq. 8.17

$$\lambda = \tfrac{27}{360}\,(20) = 1\tfrac{1}{2}$$

In Fig. 8.9, we see a high-speed, unsymmetrical cam acceleration curve with a positive acceleration period of the 4-5-6-7 polynomial.

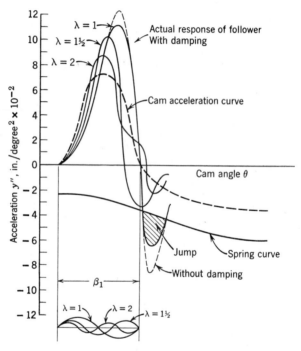

Fig. 8.9.   Jump phenomenon—transient response of follower (cam rotates 2100 rpm, $\beta_1 = 27$ degrees, and $n = 20$).

Superimposed is the actual acceleration response curve (including damping), which is found by using the methods of Art. 8.5. Also shown is the compression spring curve below the negative acceleration values to maintain constraint of the follower on the cam. Jump occurs when the response curve falls below the spring curve. Turkish[22] in his excellent article verifies this by tests. We see that jump becomes more predominant with smaller values of λ or, in other words, with more flexible systems.

A cam having a parabolic acceleration curve (not to be confused with parabolic motion) is an excellent choice. Low values of λ are allowed. Proper choice of acceleration curve shape having finite pulse values can keep jump to a minimum. It is further shown in high-speed, symmetrical or unsymmetrical acceleration curves that the vibratory jump does not exist if $\lambda > 2$. For systems having $\lambda < 2$, mathematical analysis similar to that shown in Arts. 8.5 and 8.6 is necessary to investigate the condition of jump. A direct approach for establishing the minimum allowable value of λ to prevent jump is shown by

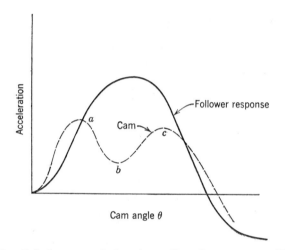

Fig. 8.10.   Polydyne cam design (note dip *a-b-c* to oppose "jump").

Kármán and Biot.[23] Increasing the spring load is a poor possibility in eliminating jump, since greater surface stresses and shorter life result. A design method for eliminating this detrimental condition with highly flexible systems is by using the polydyne procedure of the previous chapter. In Fig. 8.10, we see the highly flexible follower acceleration curve and the cam acceleration curve necessary to fulfill the polydyne requirement. Note the negative acceleration dip *a-b-c* that opposes the jump condition. It may be mentioned, however, that this dip is difficult to fabricate.

8.11 BALANCING

One of the difficulties of high-speed cam action is the unbalance of the cam mass with respect to the center of rotation. Obviously, strong vibrations are induced in the system at the cam speed fundamental harmonic. This vibration, although not completely eliminated, can be kept to a minimum by intelligent practical cam proportions, using a well-ribbed, low-mass structure (Fig. 11.20). It may be mentioned that the radial-face cam is the most balanced type of radial cam. The plain cylindrical cam is inherently completely balanced, which often justifies its adoption for high speeds.

8.12 FOLLOWER SPRING SURGE

The cam-follower compression spring has the basic function of keeping the cam and follower in contact. At high speeds, a phenomenon occurs that may seriously reduce the effective force of the spring and allow the follower to leave the cam, even though considerable surplus spring force was provided. This is called spring surge. It is a torsional wave transmitted through the wire up and down the spring at the natural frequency of the spring. It has been found to be a state of resonance of the natural frequency of the spring with the cam high-amplitude harmonics. Preventing spring surge is simple. In general, the lower the harmonic number, the higher are the vibratory amplitudes. Therefore, the natural frequency of the spring should be high enough so that if resonance occurs it will be with higher harmonic numbers and vibratory amplitudes will be kept to a minimum. The harmonic number should be 11 or higher.[22] However, a ratio as low as 9 may be used with good follower dynamics and a smooth acceleration curve. When we speak of harmonic number in this instance, we mean the ratio of the spring natural frequency to the cam speed. Equation 8.5 may be used in this analysis. However, remember that the weight of the spring substituted should be one-third the total spring weight. It may be mentioned that the phenomenon of "jump" exists for high-speed springs as well as the linkages. For further discussion of the subject of spring surge, see Jehle and Spiller.[10]

8.13 SUMMARY

Let us now conclude the chapter with a summary of both theoretical and practical design considerations for high-speed cam-follower performance.

(a) The maximum acceleration values of the cam should be as small as possible to give small inertia loads. In other words, use the maximum time and the minimum throw or movement of the follower.

(b) The pulse (da/dt) should never equal infinity because of high stresses, high resonant vibrations, and cyclical vibratory effects. That is, the cam acceleration curve should be a continuous function, preferably "smooth," giving weaker harmonics. The practical pulse values should approach that of the cycloidal curve at the dwell ends of the action. Note that a surface may appear smooth to the eye and yet have infinite pulse values. An example is the circular arc cam (Fig. 5.11) which has acceleration curve discontinuities at the intersection of the blending arcs.

(c) In general, the cycloidal cam curve will fulfill the requirements of most dwell-rise-dwell machine problems if it can be made to the required accuracy. The trapezoidal and modified trapezoidal cam acceleration curves having lower maximum acceleration values may be an improvement. All are excellent choices for low follower vibration.

(d) For highly flexible cam-follower linkages, the ratio $\lambda > 2$ eliminates "jump."

(e) Make parts as rigid as possible, i.e., keep flexibility to a minimum. Use laminated thermoset resins, carbides, and new materials if possible and practicable.

(f) Make the component moving parts of the machine as light as possible. Use new materials such as titanium, magnesium, and aluminum, if feasible. In both (e) and (f), we should aspire for a high natural frequency of the follower linkage. If resonance occurs in a given speed range, it will be with a higher harmonic number and consequently smaller amplitude harmonics.

(g) Surface finish and accuracy of fabrication are of prime importance. Be sure that cams are cut accurately in accordance with the theoretical contours so that benefit from the mathematics is not lost. Keep surface errors to a minimum. Because of surface imperfections, the actual cam dynamic characteristics rarely agree with the mathematical or theoretical ones.

(h) Backlash in parts should be held to a minimum. Preloaded bearings and followers are a possible solution.

(i) Keep the translating follower overhang to a minimum and the follower bearing guide length to a maximum. This will allow a larger pressure angle, smoother action, and a smaller cam.

(j) Use low-friction bearings, and lubricate all mating surfaces.

(k) Balance the cams with intelligent proportioning of the mass.

Furthermore, in a piece of machinery, operational difficulties may persist in spite of the application of the foregoing principles. Then, it is possible that the designed linkage and cam are improper; the conditions may be beyond the realm of a mechanical cam-follower system. It is suggested that the engineer use other systems for fulfilling the requirement, such as electrical solenoid, hydraulic, or pneumatic means. Experience indicates that these are best for long-stroke action.

References

1. S. Timoshenko, *Vibration Problems in Engineering,* D. Van Nostrand Co., Third Edition, 1955.
2. D. B. Mitchell, "Tests on Dynamic Response of Cam-Follower Systems," *Mech. Eng. 72,* p. 467 (June 1950).
3. J. A. Hrones, "Analysis of Dynamic Forces in a Cam-Driven System,' *Trans. ASME 70,* p. 473 (1948).
4. D. G. Anderson, Cam Dynamics, *Prod. Eng. 24,* p. 170 (Oct. 1953).
5. A. F. Gagne, Jr., "Design of High-Speed Cams," *Mach. Des. 22,* p. 108 (July 1950).
6. C. N. Neklutin, "Designing Cams for Controlled Inertia and Vibration," *Mach. Des. 24,* p. 143 (June 1952).
7. F. R. E. Crossley, *Dynamics in Machines,* First Ed., Ronald Press, New York, p. 403, 1954.
8. S. Timoshenko and D. H. Young, *Advanced Dynamics,* McGraw-Hill, New York, First Edition, p. 54 (1948).
9. P. Barkan, "Comment on Cam-Follower Systems article by D. B. Mitchell,"[2] *Mech. Eng. 72,* p. 1009 (Dec. 1950).
10. F. Jehle and W. R. Spiller, "Idiosyncrasies of Valve Mechanisms and Their Causes," *Trans. SAE 24,* p. 197 (1929).
11. K. Schlaefke, "Relation between Form, Frequency Spectrum and Manufacturing Accuracy of Cams," *Jahre & A. A. Luft Forschg. 19,* p. 353 (1943).
12. M. C. Turkish, "Inspection of Cam Contours by the Electronic Method," *Eaton Forum, Eaton Mfg. Co., Detroit, Mich., XI,* 1 (March, 1950).
13. P. Barkan, "Calculations of High-Speed Valve Motion with Flexible Overhead Linkage," *Trans. SAE 61,* p. 687 (1953).
14. R. P. Horan, "Overhead Valve Gear Problems," *Trans. SAE 61,* p. 678 (1953).
15. R. A. Roggenbuck, "Designing the Cam Profile for Low Vibration at High Speeds," *Trans. SAE 61,* p. 701 (1953).
16. E. H. Omstead and E. A. Taylor, "Poppet Valve Dynamics," *Journal of Aeronautical Sciences 6,* p. 370 (1938).
17. T. R. Thoren, H. H. Engemann, and D. A. Stoddart, "Cam Design As Related to Valve Train Dynamics," *Trans. SAE 6,* p. 1 (Jan. 1952).
18. E. Fuhrman, "Cams and Followers for High-Speed Internal Combustion Engines," *Engineers' Digest 12,* p. 340 (Oct. 1951); p. 368 (Nov. 1951).
19. R. D. Rumsey, "Redesign for Higher Speed," *Mach. Des. 23,* p. 121 (April 1951).
20. E. A. Edwards, "Analysis of Mechanisms with High-Speed Photography," *Mach. Des. 26,* p. 206 (Dec. 1954).
21. R. D. Hawkins, "High Speed Photography," *Mach. Des. 27,* p. 214 (April 1955).
22. M. C. Turkish, "The Relationship of Valve Spring Design to Valve Gear Dynamics and Hydraulic Lifter Pump-up," *Trans. SAE 61,* p. 707 (1953).
23. T. von Kármán and M. A. Biot, *Mathematical Methods in Engineering,* First Ed., McGraw-Hill, New York, Chapter 10, 1940.
24. F. I. Barratta and J. I. Blum, "When Will a Cam Follower Jump?" *Prod. Eng. 25,* p. 156 (July 1954).

Force Analysis

9.1 INTRODUCTION

In this chapter we shall discuss forces, loads, torques, friction, and inertia, information which is needed primarily to determine the sizes of members, the choice and life of materials, the spring size, the bearings, ultimate speeds, and the power consumption. Cams are designed to fulfill either of two requirements. The first is to give the exact location of a controlled event in a low-mass, low-torque system. An example of this type is the computer mechanism. The second is to move a high mass sometimes with a high speed. Here control of the event is usually of secondary consideration, and the system requires high torques. Thus it is necessary to determine the power consumption and prime mover size (usually a motor). The indexing or dial-feed mechanism on printing or paper-making machinery is an example of this type.

The cam-follower forces are: the static force due to the external load on the cam, the inertia force due to the follower acceleration, the acceleration force which produces vibrations, the spring force, the frictional forces, etc. In the following articles, we shall present a concise review of these forces and apply them to cam design. It is suggested that the reader refer to a good book on mechanics for elaboration.

9.2 STATIC FORCES

Static forces are gradually applied forces, and they are the principal forces in a slowly moving cam. The static-force analysis of a cam and translating follower mechanism was presented in Arts. 3.4, 3.5, and 3.6, together with the pressure-angle limitation. The examination of any system should be made in the same manner in which

255

the equilibrium conditions are:

$$\sum \text{Forces in } x, y, \text{ and } z \text{ directions} = 0$$

$$\sum \text{Moments about any point} = 0$$

9.3 INERTIA FORCES

Any body having an external unbalanced force acting on its mass m will be accelerated. A *translating body* will resist this acceleration with a reaction or inertia force

$$F_i = \frac{w}{12g}\, a \qquad (9.1)$$

where F_i = inertia force, lb.
a = acceleration, in./sec^2.
w = equivalent weight (see Art. 7.6 for elaboration), lb.

The inertia force, passing through the center of gravity of the body, has a direction opposite that of the acceleration. By D'Alembert's principle, we may make a free-body diagram of all forces and analyze the dynamic condition as a static problem (Fig. 9.1a).

For *rotating bodies,* the analysis is similar. If the body has an external unbalanced torque, it will have an angular acceleration which will be resisted by a torque reaction. The direction of this torque will be opposite to the direction of acceleration (Fig. 9.1b). The inertia reaction of an accelerated rotating body

$$T_i = I\alpha, \text{ lb-in.} \qquad (9.2)$$

where I = moment of inertia of body about the center of rotation, lb-in.-sec^2.
 = $m\bar{k}^2$.
$\bar{k}$ = radius of gyration, in.
α = angular acceleration, radians/sec^2.

9.4 VIBRATORY FORCES

In Art. 8.3, it was shown that a large pulse induces undesirable vibrations, which occur at the natural frequency of any system. These vibrations in turn produce stresses and forces which are super-imposed on the inertia force F_i. The ratio of maximum actual force or stress respectively compared to the inertia force or stress in Art. 8.3 was called the dynamic magnification factor D. The following safe values should be used in designing high-speed cams: For any curve having an infinite pulse, use $D = 2$ for relatively rigid follower members and $D > 2$ for a very flexible follower. Examples are the simple harmonic and parabolic motion dwell-rise-dwell curves. For cycloidal and similar curves with comparative maximum finite pulse values and for polydyne cams at the designed speed, use $D = 1.1$.

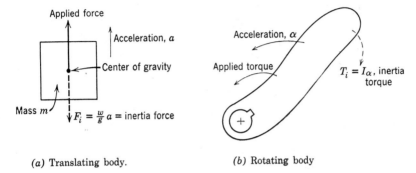

(a) Translating body.   (b) Rotating body

Fig. 9.1.   Free-body diagrams of accelerated bodies by D'Alembert's principle.

These values are based on the condition that wear, errors, and inaccuracies in manufacturing produce vibrations in spite of the exactness of the mathematical design. It will be shown in Chapter 10 that a small amount of surface waviness, surface error, or backlash in the linkage may produce high vibratory and dynamic forces. These may exceed all other forces. It is suggested that a smooth, accurate cam profile be used to gain full advantage of the mathematical qualities of the curve. In general, for rigid, low-speed members with limited backlash, vibratory forces will be negligible.

Let us remember that the cam should not run at resonance and should have a minimum of backlash to give reduced cross-over shock. Note that the resulting vibratory forces and stresses are superimposed on all others calculated. Furthermore the torque curve of a high-speed cam-follower system does not include these vibratory forces because the vibratory forces have a zero mean value transmitted to the input torque of the driver.

9.5 FRICTIONAL FORCES

Frictional resistance opposing the relative movement of contacting parts always exists in a machine. In frictional analysis, we must consider both the static and the kinetic values of friction. The best and most accurate way to include friction in the design is to measure it from the actual machine or prototype. The second approach is to use values from general test data in handbooks. A third approach is empirical and takes the friction as a percentage of the inertia torque. Friction can rarely be neglected. It is of special importance in large, heavy machinery, such as barrel indexing cams. Articles 3.4, 3.5, and 3.6 showed the frictional components on a cam driven follower. The kinetic coefficient of friction for bronze on steel equals approximately 0.10. The static coefficient of friction for bronze on steel equals approximately 0.15. Other values may be found in any mechanical engineering handbook.

If a translating follower has some flexibility and noticeable backlash, jamming in the follower guides is possible, in which case it is suggested that all friction coefficients be increased by at least 50 percent.

9.6 SPRINGS

In cam-follower systems, the follower must be held in contact with the cam at all times. This is accomplished by a positive-drive or a compression spring. This spring generally is of a helical design. The primary function of the spring force is to counteract the inertia of the follower at high speeds and to prevent the follower from lifting off the cam. It is the negative acceleration period that causes this trouble.

In Fig. 9.2, we see a cam and follower, with its displacement and acceleration curves. The critical design point is where the inertia reaction of the follower is a maximum and tends to eliminate contact with the cam. This point occurs at or very close to the maximum negative acceleration. The spring must be adequate to exceed the sum of all or some of the following forces; inertia, weight, damping, dynamic, external load, and friction. We have seen that the inertia force depends upon the speed and that, the higher the speed, the larger will be the spring needed. If the spring is too weak to sustain contact at higher speeds, a "bouncing" action will occur. Note that a certain small amount of set or loss in spring load will occur after a period of use in a machine. This loss is the result of plastic flow caused by repeated high stresses, especially at high temperatures.

The primary design factor in springs is the spring index or spring scale.

$$k_s = \frac{\Delta S}{\Delta \delta} \quad \text{lb/in.} \tag{9.3}$$

where ΔS = change in spring force, lb.

$\Delta \delta$ = change in spring deflection, in.

A plot of these parameters (Fig. 9.3) closely approaches a straight line, and the spring scale may be considered the slope of this curve.

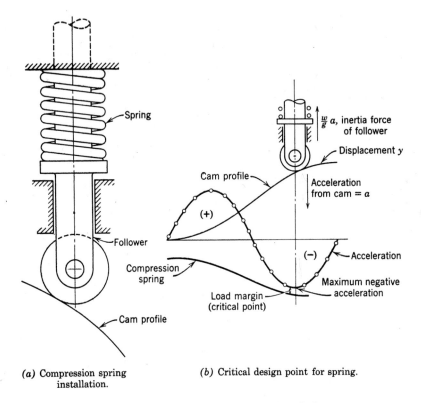

(a) Compression spring installation.

(b) Critical design point for spring.

Fig. 9.2. Follower compression spring design.

We see the initial or preload of the spring S_1 and the range of operation S_1 to S_2. S_2 is the maximum spring load. It occurs at the maximum rise point of displacement curve. Some load margin (Fig. 9.2) must be provided at the design point to allow for friction, spring surge, vibrations, and variation in spring fabrication. The spring force should exceed the net total external force by not less than 30 to 50 percent, with margins as high as 100 percent some-

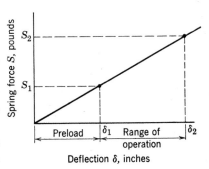

Fig. 9.3. Spring characteristic (spring scale equals slope of curve).

times used. Note that up to one-half of the lower margins cited may be directly due to spring manufacturing variations. Also, excessive spring loads are not suggested because of two primary reasons. The first is increased wear and short life of the materials. The second is that having a high pressure angle on the fall action of a cam may allow this strong spring to *drive* the cam, thus producing serious shock and noise. Backlash in the linkage obviously aids in fostering this detrimental condition.

The steps necessary in designing a spring are:

(a) Find the mass of the accelerated parts.

(b) Draw the follower acceleration curve.

(c) Convert this curve to inertia forces.

(d) Superimpose on this curve all other loads (i.e., external, frictional, weight).

(e) Draw a suitable spring characteristic curve with the minimum margin of safety for long wear life.

(f) Design the spring.

In Art. 9.9, we shall show an example of spring calculations.

9.7 TORQUE

The torque values may be required in any cam-follower system, especially for those mechanisms having heavy masses or loads. From the torque, we can determine the cam-shaft loads, the power to drive the follower, and thus the size of motor or drive. We shall see that in any system the torque is continuously changing.

Let us establish the relationship for the torque necessary for a translating roller follower, since it occurs most frequently. Let

T = torque, lb-in.

L_o = total load on the cam, i.e., inertia, weight, external, spring, friction, etc., lb.

r = radial distance to trace point, in.

α = pressure angle, degrees.

F = force normal to cam surface, lb.

In Fig. 9.4 we see the load

$$L_o = F \cos \alpha$$

Also the torque

$$T = Fr \sin \alpha$$

Substituting gives the torque at any instant

$$T = L_o r \tan \alpha \qquad (9.4)$$

But we know

$$\tan \alpha = \frac{v}{r\omega} \qquad (3.14)'$$

Substituting yields the torque

$$T = \frac{L_o v}{\omega} \tag{9.5}$$

Thus we see that the torque is proportional to the follower velocity and total load, and (in this equation) is not affected by the pressure angle. Note that at high speeds (high accelerations) the inertia load predominates in the value of the total load L_o.

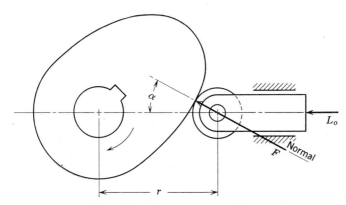

Fig. 9.4.   Cam torque analysis.

Thus, all cams discussed have a varying torque for the complete cam cycle. This is true even with a constant external load on the follower. Therefore, if these cams are driven by a motor or other power unit, the torque rating of the prime mover would have to be at least equal to the value of the maximum torque. It could be shown that a smaller torque and smaller prime mover can supply the same energy requirement by using constant-torque cams developed by trial and error. Reduced size of parts, such as gears, shafts, belts, etc., results. These constant-torque cams have found application in activating mechanisms for motor-drive spring compression, circuit breakers, and others. Similarly, a unique (hydraulically driven) inverse cylindrical cam has been applied in controlling the inclination of aircraft propeller blades as required by the changing speed of the plane. The cam profile was established to provide a constant excess of torque over a complex resistance torque curve.

9.8   DESIGN DYNAMIC LIMIT

Often the question arises in creating a new machine as to the maximum allowable follower acceleration. The author has designed and investigated a number of high-speed and high-mass machines in diversified fields and found that the inertia force (F_i, eq. 9.1) lies between certain values. That is, if F_i is kept between 500 lb and 2000 lb, we have an approximate dynamic limit for all sorts of machines. Stated differently

$$500 < wG < 2000 \qquad (9.6)$$

where G = maximum follower acceleration in g's or number of 32.2 ft/sec^2.

Generally the lower number applies to high-speed, low-mass systems (down to w equal to 5 lb) with the upper number for low-speed, high-mass systems up to w equal to 1000 lb. No explanation will be offered for this empirical relationship in terms of the many interrelated parameters, although the discussion of "jump" in Art. 8.10 indicates the necessity for this range with high-speed, low-weight linkages. Equation 9.6 is the product of the highest standards of material choice, fabrication accuracy, and theoretical design that cost limitations permit.

It should be noted that the acceleration is proportional to the square of the cam speed. Therefore for the same machine twice the cam speed means four times the dynamic limit, wG. Of course, *functional* requirements of a specific machine may require further reduction in speed and the dynamic limit. For example, certain shoe and textile machinery manufacturers having particular requirements use a value of 500 for their high-speed equipment. There are some machines in operation having wG limits as low as 250 lb and others as high as 3000 lb. In spite of the ranges and qualifications indicated, this dynamic relationship should prove an excellent guide to the designer, especially one developing a new machine. See the next article for application.

9.9 DESIGN EXAMPLE

Let us analyze the spring and cam load for a high-rotational-speed cam. As with many engineering problems, the final spring size is an intelligent compromise of many factors, e.g., cam size, pressure angle, cam contour, cam speed, surface stresses, and torque.

Example

We are given a helical-spring-loaded cam which rotates at 1200 rpm with a translating roller follower. The action is: (a) the follower rises, $1\frac{1}{4}$ in. in 160 degrees of cam rotation; (b) a dwell for 20 degrees; (c) the fall same as the rise; (d) a dwell again for 20 degrees. Furthermore an unsymmetrical parabolic curve will be utilized having the positive acceleration three times the negative acceleration to keep the spring size and surface stresses small. The parabolic curve, which is, of course, a poor choice on the basis of vibratory affect, is presented for ease of calculations. The follower linkage weighs 2 lb, and the external force on the system is constant at 10 lb.

(a) Design the spring with a margin of safety of at least 7 lb in excess of the forces tending to eliminate constrainment of the follower on the cam. This will compensate for friction, errors, etc.

(b) Find the maximum torque to drive the cam which has a limiting pressure angle of 30 degrees.

(c) Determine the dynamic limit wG, and discuss its significance.

Solution

(a) *Spring data.* In Fig. 9.5a, we see a schematic picture of the mechanism. The time = $60/1200 = 1/20$ sec/rev. The time for total rise of $1\frac{1}{4}$ in. = $160/360 \times 1/20 = 1/45$ sec.

In Fig. 9.5b, utilizing the methods of Chapter 6, we shall divide the displacement diagram into two parts, denoted by 1 and 2. We are given that $\theta_1 + \theta_2 = 160$ degrees. Since the ratio of acceleration is $3:1$, we see that

$$\theta_1 = 40 \text{ degrees}$$

$$\theta_2 = 120 \text{ degrees}$$

The time

$$t_1 = \tfrac{40}{160}\left(\tfrac{1}{45}\right) = \tfrac{1}{180} \text{ sec}$$

$$t_2 = \tfrac{120}{160}\left(\tfrac{1}{45}\right) = \tfrac{1}{60} \text{ sec}$$

Also, we know that

$$y_1 + y_2 = 1\tfrac{1}{4}$$

Substituting eq. 2.33, using the simplified method of Art. 2.11,

$$\tfrac{1}{2}a_1 t_1^{2} + \tfrac{1}{2}a_2 t_2^{2} = 1\tfrac{1}{4}$$

$$\frac{1}{2}\,a_1\left(\frac{1}{180}\right)^{2} + \frac{1}{2}\,\frac{a_1}{3}\left(\frac{1}{60}\right)^{2} = 1\tfrac{1}{4}$$

$$a_1 = 20{,}200 \text{ in./sec}^2 \uparrow$$

$$a_2 = -6750 \text{ in./sec}^2 \downarrow$$

For clarity the arrow indicates the direction of action. Also the dis-

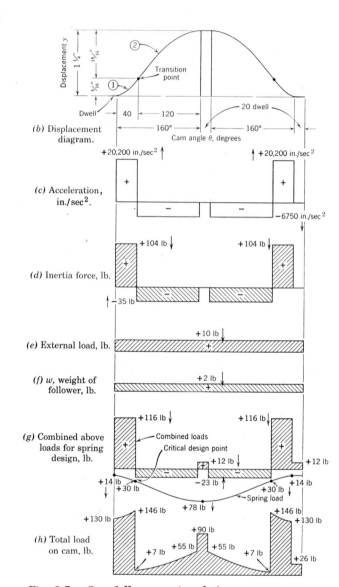

(a) Picture.

(b) Displacement diagram.

(c) Acceleration, in./sec².

(d) Inertia force, lb.

(e) External load, lb.

(f) w, weight of follower, lb.

(g) Combined above loads for spring design, lb.

(h) Total load on cam, lb.

Fig. 9.5.   Cam-follower spring design.

placements

$$y_1 = \tfrac{1}{2} \ (20,200) \ (\tfrac{1}{180})^2 = \tfrac{5}{16} \text{ in.}$$

$$y_2 = \tfrac{1}{2} \ (6750) \ (\tfrac{1}{60})^2 = \tfrac{15}{16} \text{ in.}$$

The displacement diagram is shown plotted in Fig. 9.5b. Now the inertia loads from eq. 9.1

$$F_i = \frac{w}{g} \, a$$

where w is the equivalent weight from Art. 7.6.

$$w = \frac{\text{weight of spring}}{3} + \text{weight of linkage}$$

$$= \text{assumed negligible} + 2 = 2 \text{ lb}$$

The inertia force which increases the cam surface load

$$F_i = \tfrac{2}{386} \ (20,200) = 104 \text{ lb} \downarrow$$

The critical inertia force tending to remove follower from cam

$$F_i = \tfrac{2}{386} \ (-6750) = -35 \text{ lb} \uparrow$$

The acceleration curves and inertia forces are shown in Figs. 9.5c and 9.5d. The external load on the follower is $+10$ lb $\downarrow$ in Fig. 9.5e. The weight of follower linkage is $+2$ lb $\downarrow$, Fig. 9.5f. Now, we shall combine these forces. Figure 9.5g shows the superimposed values to give a combined force diagram. We see a fluctuating load from $+116$ lb to -23 lb, etc. It was stated that the spring must exceed the net negative ($\uparrow$) loads by a margin of 7 lb. In Fig. 9.5g, we shall plot the spring load curve. At the transition point, the spring force must equal 23 lb $+$ 7 lb $= 30$ lb. At the lowest point, let us assume an initial preload of spring of 14 lb. Therefore this spring from eq. 9.3 has a spring index

$$k_s = \frac{\Delta S}{\Delta \delta} = \frac{30 - 14}{\frac{5}{16}} = 51.2 \text{ lb/in.}$$

Let us substitute to find the spring force at the maximum rise

$$\frac{S_2 - 14}{1\frac{1}{4}} = 51.2$$

$$S_2 = 78.0 \text{ lb}$$

Superimposing the spring force in Fig. 9.5g, we see that the spring force curve rises appreciably as the cam rises. Note that the spring force is $+(\downarrow)$ although it is now plotted in reverse for better understanding of margin of safety. Also, this spring may have to be redesigned to fit the mechanism or to eliminate any spring surge. Lastly, it must be remembered that the highest cam speed requires the largest spring since at any lower speed the inertia forces are reduced.

Let us now combine all the forces. This is shown in Fig. 9.5h, in which we see a fluctuating load curve. The critical load point is the transition point where the minimum difference of 7 lb exists between the spring load and the external load. A slightly stronger spring could have been chosen, with a negligible effect on the system.

It should be noted that the foregoing problem serves primarily as a guide since recalculations are necessary. One of the factors for redesign would be to include the spring weight in the equivalent weight of the follower. As a means of improving this design, we could modify the following:

(1) Change spring scale to give a smaller spring size. In this design this change would modify the initial load of 14 lb without reducing the transition point load of 30 lb.

(2) Use a larger cam angle for the action.

(3) Lower the cam speed.

(4) Change acceleration ratio and values.

(5) Use a better cam acceleration curve.

(b) *Maximum torque.* To find the maximum torque, we may first determine the cam size. From eq. 3.19, the pitch-circle radius, using the positive acceleration portion

$$R_p = \frac{fh}{\beta}$$

$$= \frac{(3.46)(\frac{5}{16})}{40\pi/180} = 1.55 \text{ in. (at the transition point)}$$

Assuming that this cam is large enough for the shaft, hub, and roller-follower diameter, the torque at the transition point by eq. 9.4

$$T = L_o r \tan \alpha$$
$$= 146 (1.55) \tan 30 = 131 \text{ lb-in.}$$

Further observation of the total load curve (Fig. 9.5h) shows that this is the maximum load value and therefore the maximum torque. The torque at the dwell points equals zero since the pressure angle is zero in magnitude.

(c) *Dynamic limit.* From earlier in the solution, we found a design dynamic limit of 104 lb. Since this is much less than the lower limit of 500 lb suggested in eq. 9.6, it is reasonable to increase the cam speed or follower linkage mass without affecting the performance of the system. The cycloidal curve is a further suggested refinement.

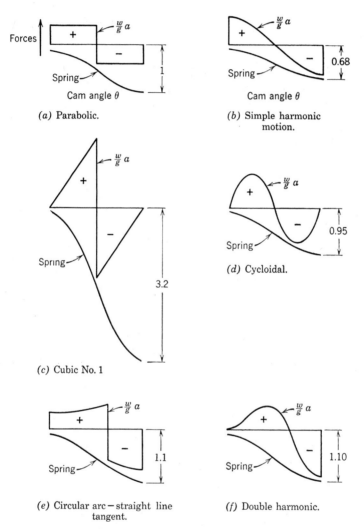

(a) Parabolic.

(b) Simple harmonic motion.

(c) Cubic No. 1

(d) Cycloidal.

(e) Circular arc — straight line tangent.

(f) Double harmonic.

Fig. 9.6.   Spring comparison of basic curves.

9.10 SPRING COMPARISONS OF CURVES

Let us compare spring sizes of some curves, using as a unit size the spring required to keep the roller follower in contact with a cam having a parabolic curve. We shall appraise the following curves, all of which operate under the same conditions: parabolic (unit size), cubic no. 1, circular arc-tangent straight-line cam, simple harmonic motion, double harmonic, and cycloidal. These curves have a 1-in. lift in 120 degrees of cam rotation. The base-circle radius is 2.8 in. The circular arc-tangent cam has one additional dimension—the distance between the cam center and the nose equals 2 in. The moving parts weigh 1 lb, and the spring has an initial load of 20 lb at zero lift. For further information, the reader is referred to the excellent discussion by Jennings.[1,2]

In Fig. 9.6, we see that the trigonometric curves, as in all other comparisons, are much better than the three remaining curves. For example, the simple harmonic motion curve has a spring size and spring force that is 68 percent of the parabolic curve values. Furthermore, the discontinuity of the acceleration curves of Figs. 9.6a, c, and e requires a much stronger spring than indicated in these comparisons. Vibratory follower response discussed in Art. 8.3 is the cause. Although the flat-faced follower is not compared, it has an excellent characteristic of small spring size.

References

1. J. Jennings, "Calculating Springs for Cams," *Mach., London 57*, p. 433 (Jan. 16, 1941).
2. J. Jennings, "High-Speed Cam Profiles," *Mach., London 52*, p. 521 (July 28. 1938).

C H A P T E R   T E N

Surface Materials, Stresses,

and Accuracy

10.1

In this chapter we shall consider the mating follower and cam surfaces. The practical choice of materials, surface treatment, surface finish and surface stresses, lubrication, accuracy, and other factors will be discussed. We must recognize that much on the subject of surfaces in contact is controversial and experts disagree as to the phenomena of surface stress and life. Furthermore, the performance of cam-follower systems tested in the laboratory under controlled conditions differs from the performance in the field for two primary reasons: first, the true dynamic and operating loads are never known under operation; second, the machined surface may be very different from the theoretical mathematically established contour. For these reasons, very little information on this subject is correlated or organized. However, in this chapter we shall attempt to compile the available data; it should be used with discretion. Experience and performance in the field is the final criterion in all designs.

Materials and Stresses

10.2 SURFACE

Let us examine the appearance and quality of a surface prior to our study of the complex action of contact. Theoretically, surfaces are represented as being smooth and flat. However, the actual surface does not present perfect smoothness. It is a three-dimensional boundary with irregularities and sometimes ragged edges, varying from sharp peaks to small bumps (Fig. 10.1).

Other factors for consideration in the determination of surface value are hardness, material, and metallurgical structure. It is known that the surface quality after machining is almost always changed by the destructive action of metal removal, especially in grinding. The metal is affected to an appreciable distance below the surface.

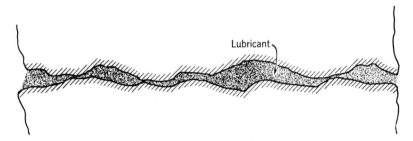

Fig. 10.1. Surfaces in contact.

The effects on the surface are high stresses due to cold working, higher susceptibility to corrosive action, and metallurgical changes due to high cutting temperatures. In general, the surface change is related to the pressure and the speed of the metal-finishing operation. Thus even to define a surface is a complex problem.

10.3 SURFACE COMPRESSIVE STRESS—CYLINDRICAL PINS AND ROLLERS, CURVED-FACED AND FLAT-FACED FOLLOWERS

The state of deformation and stress existing between two elastic bodies in contact under load was established in the classic work by Hertz.[1] Let us consider two cylinders in contact. Theoretically, if the bodies were perfectly rigid, they would have straight-line contact. Actually the bodies under load deflect to give area contact (Fig. 10.2). The amount of deflection or strain (proportional to the stress) depends largely on the modulus of elasticity.

Hertz indicated the stress distribution of two contacting cylinders, as shown in Fig. 10.3a. Assuming perfect alignment of the contacting bodies, we see that the surface compressive stress is a maximum at the center of the contact area and decreases to zero at the ends. The area of contact of the two cylinders is rectangular in shape. However, actually the deflection and misalignment of the cam shaft will cause increased cam stresses. In this respect, cams have much in common with gears in that we do not know the actual distribution of the load over the cam surface, particularly when the followers present plane surfaces to the cams. Figure 10.3b shows this extreme stress condition, which is caused by limiting the lengths of the contact line and contact area. Some manufacturers somewhat alleviate this condition in roller followers, by grinding a crown on the roller surface.

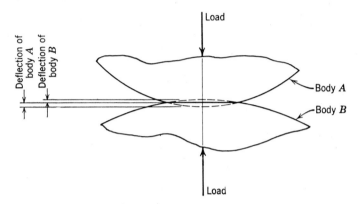

Fig. 10.2. Deflection of curved surfaces in contact.

The automobile cam has a taper ground on its surface (of a few thousandths) in the cam width to give uniform stress distribution by compensating for the cam shaft deflection.

Two cylinders in contact and in alignment produce a maximum compressive stress below the surface

$$S_c = 0.57 \left[\frac{F\left(\dfrac{1}{\rho_c} + \dfrac{1}{\rho_f}\right)}{t_h\left(\dfrac{1 - \mu_c^2}{E_c} + \dfrac{1 - \mu_f^2}{E_f}\right)} \right]^{\frac{1}{2}} \tag{10.1}$$

where S_c = maximum compressive stress at any point, $lb/in.^2$
  F = normal load on cam profile, lb.
  ρ_c and ρ_f = radii of curvature of cam and follower, respectively, in.
  t_h = thickness of contacting cam and follower, in.
  μ_c and μ_f = Poisson's ratio for cam and follower, respectively, having average values: steel = 0.30; cast iron = 0.27; bronze = 0.34.
  E_c and E_f = moduli of elasticity of cam and follower, respectively, $lb/in.^2$ Average values are: steel, 30,000,000; cast iron, 12,000,000 to 23,000,000, depending on class; brass or bronze, 15,000,000; nylon, 400,000.

We note in Fig. 10.4 that the radius of curvature is positive when the profile is convex and negative when the profile is concave. For a flat-faced follower, substitute in eq. 10.1 its radius of curvature, ρ_f, equal to infinity. The radius of curvature of any cam may be found by measurement from the cam layout (generally giving a reasonable

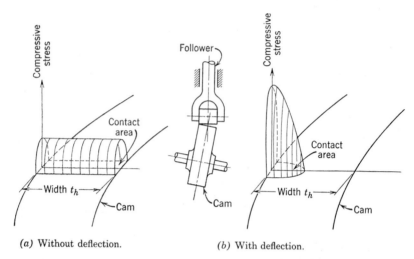

(a) Without deflection.　　　　　　(b) With deflection.

Fig. 10.3.　Stress distribution on cam—cylindrical-, roller-, pin-, curved-, and flat-faced followers.

precision), or by the exact mathematical method described in Chapter 3.

The value of the thickness t_h depends upon the rigidity of the cam shaft and cam since the cam shaft deflection and misalignment

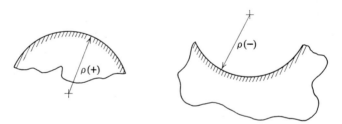

Fig. 10.4.　Radius of curvature.

previously discussed may give a poor stress distribution. For automotive application[2] the stress on the automotive cam cannot be carried in any width exceeding $\frac{1}{2}$ in. For a similar relatively flexible cam shaft, the value of $t_h \lesssim \frac{1}{2}''$ should be applied to eq. 10.1. The designer should keep this in mind and decide according to the rigidity of his system.

At any cam speed, the load and cam radius of curvature are

changing along the profile. Thus the stress is different at every point on the cam. Theoretically, all points on the cam should be investigated for maximum stress. Usually, the nose of the cam with its small radius of curvature presents the largest compressive stress. Also, it was shown in Chapter 3 that the smallest cam radius of curvature can be increased by reducing the maximum negative acceleration. Thus, an unsymmetrical cam acceleration curve having the positive acceleration greater than the negative acceleration will reduce the nose stresses and the size of the compression spring needed.

Example

A tangent cam has a nose radius of $\frac{1}{4}$ in. At 1200 rpm, a maximum force of 132 lb acts at the flank (at the transition point in the displacement curve). The force on the cam nose is 150 lb, the pressure angle at the flank point is 20 degrees, and the roller follower is $\frac{3}{4}$ in. in diameter. The cam thickness is $\frac{3}{4}$ in. Both the cam and the follower are made of steel. The cam shaft and bearing supports are relatively rigid. Find the maximum compressive stress for the nose and the flank at this speed.

Solution

The force distribution gives the normal force on the cam flank surface

$$F = \frac{132}{\cos 20} = 140 \text{ lb}$$

From eq. 10.1, the maximum compressive stress

$$S_c = 0.57 \left[\frac{F\left(\dfrac{1}{\rho_c} + \dfrac{1}{\rho_f}\right)}{t_h \left(\dfrac{1 - \mu_c^2}{E_c} + \dfrac{1 - \mu_f^2}{E_f}\right)} \right]^{\frac{1}{2}}$$

Substituting for the flank compressive stress,

$$S_c = 0.57 \left[\frac{140\left(\dfrac{1}{3/8} + \dfrac{1}{\infty}\right)}{\dfrac{3}{4}\left(\dfrac{1 - 0.3^2}{30 \times 10^6} + \dfrac{1 - 0.3^2}{30 \times 10^6}\right)} \right]^{\frac{1}{2}} = 51{,}800 \text{ lb/in.}^2$$

The nose stress

$$S_c = 0.57 \left[\frac{150\left(\dfrac{1}{\frac{3}{8}} + \dfrac{1}{\frac{1}{4}}\right)}{\dfrac{3}{4}\left(\dfrac{1 - 0.3^2}{30 \times 10^6} + \dfrac{1 - 0.3^2}{30 \times 10^6}\right)} \right]^{\frac{1}{2}} = 84{,}800 \text{ lb/in.}^2$$

We see that the nose is subjected to a larger stress than the flank, the nose radius of curvature being the most significant factor.

10.4 SURFACE COMPRESSIVE STRESS—SPHERICAL- AND CROWN-FACED FOLLOWERS

Now let us present a method for finding the surface compressive stress of a spherical-faced follower. Within reasonable accuracy the relationships may also be applied to the crowned roller follower.

In Fig. 10.5a, we see the ideal stress distribution in which the load or stress on the follower is symmetrical with the cam. The

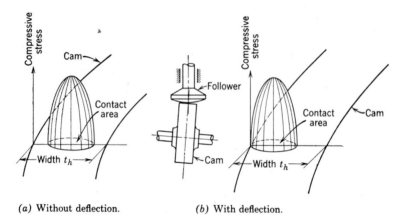

(a) Without deflection.   (b) With deflection.

Fig. 10.5.   Stress distribution on cam—spherical-faced follower.

area of contact has an elliptical shape. In Fig. 10.5b, the stress distribution of an actual cam with extreme misalignment and deflection is shown. We see that the same stress values exist which are shifted from the deflected side of the cam. This is the primary advantage of a spherical follower as compared to the serious stress increase exhibited with the cylindrical roller or flat-faced follower. The spherical radii used are from 10 to 300 in.

The following equations were obtained from references.[2,3]   The maximum compressive stress

$$S_c = \frac{0.0469 F^{1/3}}{\left(1 + \dfrac{\rho_f}{\rho_c}\right)^{0.237}} \left(\frac{\dfrac{2}{\rho_f} + \dfrac{1}{\rho_c}}{\dfrac{1 - \mu_c^2}{E_c} + \dfrac{1 - \mu_f^2}{E_f}}\right)^{2/3} \qquad (10.2)$$

Example

Let us repeat the example of Art. 10.3 for a 100-in.-radius spherical-faced follower instead of a roller. A tangent cam has a nose radius of $\frac{1}{4}$ in. At 1200 rpm the force on the cam nose is 150 lb. The cam thickness is $\frac{3}{4}$ in. Both the cam and the follower are made of steel. Find the maximum compressive stress on the nose of the cam.

Solution

From eq. 10.2 we find the stress

$$S_c = \frac{0.0469(150)^{\frac{1}{3}}}{\left(1 + \frac{100}{\frac{1}{4}}\right)^{0.237}} \left(\frac{\frac{2}{100} + \frac{1}{\frac{1}{4}}}{\frac{1 - 0.3^2}{30 \times 10^6} + \frac{1 - 0.3^2}{30 \times 10^6}}\right)^{\frac{2}{3}}$$

$$= 124{,}500 \text{ lb/in.}^2$$

It should be noted that the actual stress would be slightly higher because of the additional tangential stress, induced by the sliding frictional force component on the contacting surfaces.

10.5 WEAR PHENOMENA

The action between two surfaces in contact is a difficult problem involving the statistical relationship between many interrelated variables. Among the parameters are the surface roughness, waviness and stresses, prior history of machining, moduli of elasticity, friction (rolling and sliding), materials, lubrication, corrosion, and dynamic loads. We are concerned with wear, which in a broad sense is related to the friction between the two surfaces. Almen[5] gives us the following excellent categorization of wear:

1. Removal of metal. This is caused by (a) *abrasive* in the lubricant; (b) *corrosion;* (c) *tearing away* of high surface points; (d) *fatigue,* which is commonly called *pitting.*

2. Transfer of metal between surfaces. This occurs with sliding and produces fusion of contacting micro-points under high temperature, which is called *welding, scoring, galling, scuffing,* and *freezing.*

3. Displacement of metal by *plastic flow, superficial wear,* or *smearing.* This occurs in all metals in contact under high pressure.

In cams and followers, all three forms of wear may occur at one time.

Abrasive is a grit or removed metal particles, carried in the lubricant, which produce scratches in the surface. These scratches are usually serious. The best solution to this problem is to remove the source of this abrasive material by filtering the oil, filtering the air, or reducing the load to eliminate the production of destructive metal particles. Sometimes using a more viscous lubricant and a detergent may reduce this type of wear.

Corrosion can be eliminated by choosing proper materials and lubricant and by keeping the moisture content to a minimum.

Fatigue, or *pitting,* as it is commonly called, is produced by a repeated high compressive load on the surface of the softer of the two metals in contact. This continual alternate flexing exceeds the fatigue limit of the material, so that the high spot flakes out leaving a pit behind. Sometimes oil under high hydrostatic pressures is trapped in this hole, furthering the pit and forming a crack. Rough surfaces, high sliding velocities, corrosion, grinding cracks, thin layers of casehardening, high dynamic loads, and surface checks from heat treatment accelerate this action. The micromechanism of surface fatigue failure is unknown, although it is recognized as originating on steel surfaces or below the surface of chilled cast iron, the latter having a low shear fatigue strength. In general, a close-grain structure gives optimum resistance to surface fatigue. Note that shot-peening is employed to increase the fatigue life of cams.

For the action of removing metal by *welding, scoring,* and *galling,* the reader is again referred to Fig. 10.1. There we see that the actual contacting surfaces consist of minute high points. These points of small area are subjected to high pressure, regardless of the load on the surfaces. Instantaneously high temperatures are produced when the surfaces slide on one another. Fusion of the points results. Wear is produced when further applied force in the direction of sliding breaks the metal at the junction points. The softer metal is removed, leaving scratches and grooves sometimes adhering to the other surface. Scoring is due to the action of pressure, sliding velocity, and the resulting instantaneous temperatures.

In the last wear condition, the metal high points are reduced by *plastic flow,* sometimes called superficial wear or smearing. Under this action, the metal is displaced into the low points or valleys of the surface, which may occur quickly under high loads (high frictional heat) or slowly under low loads. Wear life of machine parts is often increased by properly controlled initial plastic flow. This provides smoother surfaces and cold working or strengthening of the softer materials. In machinery, controlled superficial wear is called "run-in."

The most important yet least understood factor in all wear study is the effect of dissimilar metals on each other. Tests show that some combinations of metals are compatible for long wear life whereas others are poor. Table 10.1[6] indicates some interesting metal combinations. We must realize that hardening the material is successful in increasing wear life only to a limited degree.

Table 10.1   Metal Combinations

Good	Poor
Cast iron and phosphor bronze	Hardened steel and hardened bronze
Hardened steel and phosphor bronze	Hardened nickel steel and hardened nickel steel
Cast iron and soft steel	Soft steel and bronze
Babbitt and soft steel	Soft steel and soft steel
Soft brasses and soft steel	Soft steel and laminated thermoset resins
Hardened steel and soft bronze	Soft steel and nylon
Hardened steel and brass	
Hardened steel and cast iron	
Hardened steel and laminated thermoset resins	
Hardened steel and nylon	

Returning to cams and followers, if we compare the wearing qualities of basic contours, again we shall find that the trigonometric family, i.e., simple harmonic motion, cycloidal, and double harmonic curves, gives longer wear life than the basic polynomial family, i.e., straight-line, parabolic, or cubic curves. This applies for both low- and high-speed action. Furthermore, offsetting the mushroom follower, as shown in Art. 5.25, gives a corkscrew motion to the follower and distributes wear.

10.6 CHOICE OF MATERIALS

Selecting materials to fulfill the requirements of smooth cam action with minimum wear over a long period of time is governed primarily by experience. Most engineers prefer to make the roller of softer or first-destroyed material for easy replacement, although there are installations in which the cam is of the softer material and has replaceable cam tracks. Roller follower and cam surface failures are primarily fatigue failures with some galling or welding wear evident. As the percent of rolling increases, the amount of scuffing wear decreases. Thus, spherical- curved- or flat-faced followers with pure sliding basically have galling wear action.

Table 10.2 Choice of Materials for Cams and Followers*

Follower Surface Material	Cam Surface Material	Pure Rolling† Allowable Compressive Stresses, S_c, psi (Eq. 10.1)	Wear Life Resistance, 100% Sliding	General Comments
Hardened and ground steel, Rock. C 51–52	**CAST IRON** Gray cast iron	55,000	—	These cast-iron cams are very good sliding wear-resisting materials and inexpensive.
	Chilled cast iron	—	Excellent; automobile field suggests $S_c = 190{,}000$ in eq. 10.2	
	Nickel cast iron—as cast	66,000		
	Nickel cast iron—heat-treated, Rock. C 35–40	79,000	Excellent	
	Molybdenum cast iron—as cast	82,000		
	Molybdenum cast iron—hot quench treatment	113,000	—	
	Meehanite cast iron—heat-treated	80,000	Excellent	Better than above; good shock resistance, vibratory damping and reduced noise.
	Ductile cast iron	—		
STEEL	SAE 1010—induction hardened, Rock. C-60	250,000		
	SAE 2515—ground and lapped, case-hardened, 0.040″ depth	250,000	—	
	SAE 4615 Carburized	Excellent		Hardened to give good wear life and tough core for extreme shock. Commercial antifriction bearing and roller follower outer races.
	SAE 8620 Carburized	Excellent		
	SAE 52100 Rock. C 60–63	Excellent		
1.05% Carbon tool steel, Rock. C 60–63	Low-carbon free machine steel case-hardened, 0.045″ depth, Rock. C 56–61	250,000		
	SAE 4340, induction hardened, Rock. C 48–50, 0.045″ depth	250,000	Good	
Hardened tool steel	Medium-carbon low alloy (0.47% C, 1% Mn, 0.65% Cr, 0.2% Mo.) machine steel, Rock. C-30, ground	230,000		
	Same plus Parco lubrite added	290,000		Parco lubrite increases wear life.
Hardened and ground steel	Nitralloy	Good		Good shock resistance.
	Tungsten carbide or stellite	—	Best of all	Sometimes welded at critical points in cam contour. For severe sliding wear but expensive.

STEEL	BRONZE			
Hardened and ground steel, Rock. C 51–52	Phosphor bronze—SAE 65, 80 BHN, chill cast / Nickel bronze—80 BHN	82,000 / 76,000	Good	Smooth action, long life, excellent shock resistance, vibratory resistance and noise, low modulus of elasticity; minimizes effects of (1) inaccuracies in contour, (2) and impact especially at cross-over point, and (3) inertia due to start and stop. Used with combinations having rolling and sliding.
	Miscellaneous bronze and copper, Ampco 18, 20, Bosworth 41, hardened beryllium copper, Mueller bronze	—		
Nickel cast iron Hot quench (CAST IRON)	Phosphor bronze SAE 65	83,000	—	—
Cast iron (IRON)	Cast iron	—	Excellent	
Hardened and ground steel (MISCELLANEOUS)	Laminated thermoset resins	—	Poor	Excellent noise reduction. Used in high-speed textile machinery in oil coolant bath.
	Hardened and ground steel	—		Excellent in drastic roller speed reductions, shock, wear, noise, vibrations, abrasive, and limited lubrication oil or water. Moderate loads. Sometimes noticeable wear in which mating material dimensionally may change.
Nylon (PLASTICS)	Nylon	—		Inferior to steel on nylon.

BHN = Brinell hardness number
Rock. = Rockwell " "

*Miscellaneous materials have also been used for making cams. These depend largely on fabrication methods and production. For example, sintered metal powder (hardenable) has been utilized for large production, small sizes.

† To be safe, as an approximation for 10 percent, sliding decrease values one-fifth. For 75 percent sliding, decrease values two-thirds. For shock, misalignment and dynamic loads, reduce values further.

Cam and a roller follower

The simplest approach to the complex overall problem of surface life is to make tests to determine a reasonable wear life. As mentioned before, this approach is suggested because of a lack of basic understanding of wear phenomena. Buckingham and Talbourdet[6,7,8,9] ran tests on cylindrical roller followers in contact with a cam surface under controlled conditions of rolling and alignment. They used different materials of various hardnesses and surfaces in contact and arbitrarily related the data to the Hertz formula previously shown, eq. 10.1. Table 10.2, in which allowable stresses are given for pure rolling action, itemizes the results of the study. If the Hertz compressive stress S_c is kept below the value given in Table 10.2, the life of the contacting surfaces will be more than 300×10^6 cycles. For shorter running life, the values shown may be increased. It was found that, in general, these stresses should be reduced one-fifth for 10 percent sliding and two-thirds for 75 percent sliding of the follower roller. In addition, if unusual dynamic shock load or excessive sliding, deflection, misalignment, abrasive in oil, or corrosion of surface is evident, these values must be further reduced. Furthermore, to guide the engineer, additional materials have been indicated in Table 10.2. Other materials not listed may be acceptable substitutes.

In designing rolling surfaces that are casehardened, it is sometimes found that pieces of the case chip out and leave a large, serious pit. These pits, which then obviously accelerate surface wear, are theoretically related to the shear stress resulting from surface pressure having a maximum value slightly below the surface. If this problem occurs, it is suggested that the case depth be kept twice the distance of the point of maximum shear stress.[11] In practice the case depth generally varies between 0.030 in. and 0.060 in.

Roller followers are constructed in many ways (Art. 5.24). Very often antifriction bearings are used in which the load capacity is related to both the contacting surface materials and the capacity of the bearing, i.e, rolling elements. Table 10.3 classifies the available data to aid the engineer in establishing the proper follower design and choice of materials.

Cam and a flat-faced follower

As stated previously cam-follower sliding failures are primarily of the galling type. Sliding fatigue failures are evidenced, especially in systems experiencing large dynamic or shock loads. No experimental data are available for the designer to aid him in deciding on a reasonable sliding wear load between surfaces in contact. However, the surface galling action of gear teeth[5] shows a statistical correlation indicating that the sliding velocity is an important factor. This is verified by the author's own experience.

Table 10.3   Load Design Guide for Roller Follower

Commercial Roller Follower (Needle Bearings)		Commercial Anti-friction Bearings	Roller Surface of Different Materials
With Stud, Fig. 5.28b	Without Stud, Fig. 5.28a		
Table 5.3 which is arbitrarily based on conventional bending theory of cantilever beams or the capacity of roller elements, whichever is less.	See catalogue. Also, do not exceed one-third the static capacity of the bearing (rolling elements).	See catalogue. Also, do not exceed one-third the static capacity of the bearing (rolling elements). For ball bearings, use eq. 10.1 $S_c = 80,000$ lb/in.^{2} for a hardened steel Rock. C-60 cam, and $S_c = 35,000$ lb/in.^{2} for a cast iron cam with failure occurring at approximately 3 times these values.[13]	See Table 10.2. If commercial antifriction bearings are used (Figs. 5.28c and d), compare values to the capacity of the bearing (rolling elements) whichever is smaller.

Table 10.2 lists the comparison between materials. The most widely accepted metal characteristic reducing the tendency to gall is *self-lubrication*. Self-lubrication is exemplified by porous powder metals (oil-impregnated), gray cast iron, Meehanite, ductile cast iron, graphitic steels (graphite-lubricated), and leaded bronze (lead-lubricated). These materials suffer less damage when galling takes place. Other materials with much sliding and high loads may have their wear life increased by the frequent addition of phosphate coatings, Ferrox, and electrographite especially during run-in periods.

10.7   SUMMARY FOR LONG MATERIAL LIFE

No explanation or theory is available for full understanding of the life of surfaces in contact. Therefore, it is no surprise that in cam-follower surface actions there is disagreement concerning service performance. However, certain basic conclusions can be presented for the proper choice of materials.

(a) Improve the cam action by improving materials only after contour choice and contour accuracy have been definitely established. In other words, the system should first be dynamically analyzed and profile cutting accuracy verified.

(b) Gray cast iron in combination with hardened steel is probably the cheapest choice for high sliding galling action. Chilled cast iron and alloyed cast iron are among the best materials.

(c) Bronze or nylon in combination with hardened steel have good wear life, especially under high shock loads. They give less noise, less vibrations, and smoother action and compensate for inaccuracies in the contour surface.

(d) Laminated thermoset resins are excellent for noise reduction. A lubricant coolant is generally necessary.

(e) For high sliding conditions (galling), Meehanite in contact with hardened steel is an excellent choice giving long life, good internal vibratory damping characteristics and easy machining.

(f) Hardened tool steel in contact with hardened steel gives the best combination under rolling action when shock and sliding action occur.

(g) Good surface finish with a lapping and polishing action is essential with hardened materials. It is not as important with ductile materials.

(h) Controlled "run-in" period under light load for a long time will further increase life.

(i) Surface contamination by the use of chemical compounds such as phosphate (Parco-Lubrite), Ferrox and electrofilm graphite aid life and wear of high sliding velocity (galling) surfaces. Sometimes the application of silicone, extreme pressure lubricants (E.P.), or molybdenum disulphide, or increasing the load capacity of lubricating oil helps.

(j) A thin lubricant (10 SSU at 100° F) is suggested for roller followers. A high-viscosity oil, grease or chemical compound is employed for the flat-faced follower sliding surfaces.

(k) High speeds generally necessitate a continuous flow of lubricant in order to reduce noise and remove heat.

Accuracy

10.8 INTRODUCTION

This subject is one of the most significant. Accuracy, cost, and time of contour fabrication are specific determining factors in the performance of a cam-follower system. Poor manufacturing techniques can seriously impede the functional ability of the mechanism. Thus the engineer must have a working knowledge of machine tools and their dimensional limits for specific manufacturing applications. This cannot be overemphasized. We must realize that cam contour manufacturing is a costly, slow, tedious job because of the random shapes of the cam and that very few cams do not have a hand operation at one

Fig. 10.6.   Some high-accuracy cams (computing mechanism).
Courtesy Ford Instrument Co., Long Island City, N.Y.

stage in their making. As emphasized throughout this book, high-speed machines and computer cams require the highest accuracy for reasonably acceptable performance. In high-speed machines, small errors in contour produce excessive noise, wear, and vibrations. Furthermore, smoothness of a cam is no indication of the inertia shocks that the profile may cause. For instance a cam comprised of circular arcs (Fig. 5.11) although smooth is not mathematically continuous at the blending points. Only by mathematical continuity of the acceleration curve can there be freedom from inertia shocks, in addition to surface precision. Figure 10.6 shows some accurate cams used in computing mechanisms.

In this chapter no attempt will be made to delve into the various methods of cam fabrication. We shall discuss accuracy of cam manufacture with the methods utilized to achieve this precision determined by experience and the availability of machine tool equipment. References at the end of the chapter will be helpful. Also, every cam developed must be individually analyzed since no unified theory exists relating the critical parameters of speeds, curve choice, and manufacturing accuracy.

10.9 CONTOUR FABRICATION METHODS

The method employed to produce a particular cam is a function of its contour, application, and quantity of production. One would certainly not manufacture a cam for a simple clamping fixture with the same degree of precision and wear characteristics as a high-speed textile machine cam. Depending on the shape and application, cams may be produced either by the machine tools found in the ordinary shop or by specially designed machines.

Cams are generally produced in the following ways: *layout cutting*, *increment cutting*, or *tracer control cutting*. Miscellaneous methods employed are flame cutting, die casting, die forging, stamping, and powder metallurgy. In addition, a *finishing* operation is often necessary for the layout and increment cutting methods.

In *layout* cutting, as the name implies, the machinist constructs the cam on the cam blank. The machine operator, by moving the proper machine tables, cutters, or workpiece follows this scribed line as accurately as possible. Sometimes a magnifying glass is employed to increase the accuracy. The equipment required includes the saw, the file, the milling machine with a rotary table, and the shaper. This method obviously produces low productions and poor precision, with the final results determined by the skill and experience of the operator. Enlarged master cams are sometimes made this way.

Increment cutting consists of fabricating the contour by intermittent cuts with a series of scallops or flats being formed. The latter are tangent to the desired cam profile. Sometimes these ridges and valleys are removed by hand operations or grinding. The angular separation of the flats or scallops governs the accuracy obtained. This increment method is used for making master cams or cams in small quantities, and the machines generally utilized are the milling machine with dividing head and the jig borer. It is a tedious and slow process providing highly accurate profiles.

In the *tracer control* cutting method, the cam surface is either milled, shaped, or ground with the cutter or grinder guided continuously by an information device, such as a hardened-surface template (master cam), or punched tape. This is the best method for producing large quantities of accurate profiles. The machines utilized are the profile miller, the die sinker, the vertical miller with cam slide attachment, and special cam milling and grinding machines. Sometimes, for greatest accuracy, both the master and production cams are cut in this manner, in which the template made by the increment cutting method is called a "leader." Figure 10.7 shows a commercial cam miller.

This tracer control cutting method has been elaborated[22,35,38] to the point of fabricating simple cam shapes without an operator. The machine consists of a computer supplying data directly to a servo-mechanism which automatically and accurately positions a continuously moving cutting tool. A feedback circuit corrects for detrimental inherent vibrations of the cutter. Information is supplied to the calculating machine on a punched tape or by other means.

10.10   SURFACE IRREGULARITIES

Some cam and follower surface imperfections (deviations from the theoretical contour) are often difficult to control or determine. Depending on their magnitude, these surface irregularities may induce shock, noise, wear, and vibrations. The vibrations produced are superimposed and may exceed all other kinds discussed throughout this book.

Three kinds of surface imperfections may be evident: *errors, waviness,* and *roughness.* It must be remembered that the effect of these three may be compounded.

Errors, local in character, may be large or small or of any shape and duration. They can be produced by an incorrect setting of the milling cutter, a file scratch, or by holding the grinding wheel at a point on the cam for too long a time. These mistakes may impair the performance of any system.

Waviness, on the other hand, is a periodic imperfection. It is a uniform distribution of high and low points of longer duration than roughness. Waviness of a surface is the vertical distance between peaks and valleys of relatively long wavelengths. It may be produced by a milling cutter taking too large a feed in the continuous generation of a cam or by the increment locations of the milling cutter in which scallops are formed that can never be eliminated. Because waviness is periodic, its successive effect may be a vibratory reinforcement at certain harmonic speeds, which may seriously impede operation.

Roughness is a random distribution of many small high and low points. Since they are random, any change is as likely to decrease

previously existent vibrations as to increase it. The net effect is
very low-level vibration (usually negligible) which, furthermore, is
absorbed by damping due to residual oil on the surface. Roughness

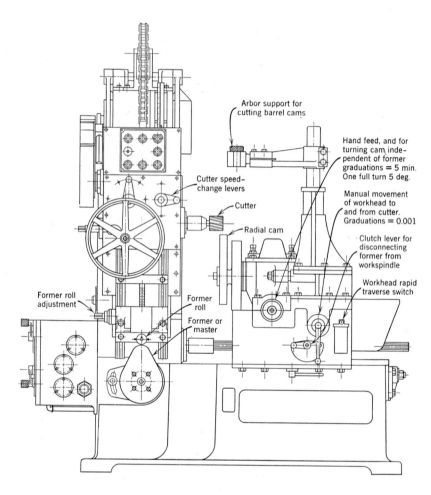

(*a*) **Cutting a radial cam.**

Fig. 10.7.   Commercial cam milling machine.

is a relatively short-wavelength irregularity. The maximum established wavelength for surface roughness is about 0.50 in. or 50,000
microin.

In practice, many cams differ in character from the theoretical
because they are difficult to form. For example, tracer-control

production grinding equipment resists a sudden change in the acceleration curve of a poorly designed master cam and thereby generally produces a better acceleration characteristic on the cam fabricated. Hence, the actual cam is an improvement over the master cam. Departure from the true dimensions is caused by the mass and elasticity of the grinding machine. Other times master cams that

(b) Cutting a cylindrical cam.

Courtesy Rowbottom Machine Co., Waterbury, Conn.

should generate perfect profiles have to be doctored to produce the best cam. Another example is the polydyne cam (Fig. 8.10) in which the sharp dip in the acceleration curve is a doubtful fabrication possibility. Furthermore, chatter, starting from a shock excitation of the discontinuous acceleration curve, can produce serious waviness in the surface being formed. Thus, *the cam fabricating machine tool is a dynamic system performing in a similar manner to the operation of the cam-follower mechanism.*

Imperfections in the cam profile always produce high-frequency and low-amplitude vibrations. The only practical possibility of improvement is more expensive, requiring more accurate fabrication and checking (preferably dynamic characteristics) of surface smoothness. Thus, a surface may appear smooth by eye or to feel and yet be made up of serious vibration-inducing imperfections. It may be dimensionally acceptable yet dynamically poor, or vice versa. Therefore, if, in an existing machine, vibration is a problem, it is suggested that the actual surface accuracy be analyzed before conclusions are drawn as to the effect of the existing cam curve. The theoretical contour and actual contour rarely agree.

10.11 INCREMENT-CUTTING ACCURACY

As stated previously, the cam and cutter sizes, and the number of machine settings (using the increment cutting method) determine the extent of filing necessary and the final accuracy of the profile. For accurate master cams, settings must often be in half-degree increments calculated to seconds. For high-speed machine cams, usual practice is to use about 1-degree increments (dependent on the size of the cam). In Chapter 4, we showed the equations for calculating the profile in which the cam dimensions are given in rectangular or polar coordinates. For reduced mathematical work, the grinder or cutter is often made the same diameter as the roller follower. Now let us determine the magnitudes of the scallops or flats to establish the degree of accuracy of this small production-run method.

Scallops (*peripheral cutting*)

In Fig. 10.8a, let O_1 and O_2 be the respective centers of the cutter for the increment angle $\Delta\theta$. Let

r_c = distance from center to cam profile, in.
R_g = radius of cutter or grinder, in.
σ = error or deviation from theoretical cam profile, in.
$\Delta\theta$ = angular increment, radians.

The circumferential distance between centers O_1 and O_2 approximately equals $r \, \Delta\theta$. In the right triangle O_1bc, we observe

$$R_g{}^2 = (bc)^2 + (O_1b)^2$$

which approximately equals

$$R_g{}^2 = (\tfrac{1}{2}r_c \, \Delta\theta)^2 + (R_g - \sigma)^2$$

Expanding and dropping σ^2 since it is negligible yields

$$\Delta\theta = \left(\frac{8R_g\sigma}{r_c^2}\right)^{\frac{1}{2}}$$ (10.3)

The more settings used, the less hand filing and hand polishing are needed. In some cams, we can mill equal settings throughout but as

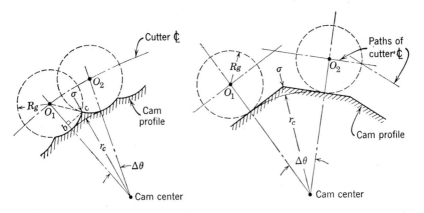

(a) Scallops—peripheral cutting.   (b) Flats—tangent surface cutting.

Fig. 10.8.   Increment cutter waviness.

radius r_c becomes smaller and we approach the center of the cam the increments $\Delta\theta$ may be increased without loss in accuracy.

Flats (*tangent surface cutting*)

Now let us establish the cam accuracy with the cutter or grinder moving tangent to the cam at each increment angle position. In Fig. 10.8b, we see

$$\Delta\theta = 2\cos^{-1}\left(\frac{r_c}{r_c + \sigma}\right)$$ (10.4)

Example

Suppose that we are to cut a cam by the peripheral incremental method, using a ¾-in.-diameter cutter. The radial distance of the cutter from the cam center averages about 3½ in. What angular spacing should be established with the scallop height held to less than 0.001 in.?

Solution

From eq. 10.3, the increment angle

$$\Delta\theta = \left(\frac{8R_{g}\sigma}{r_c{}^2}\right)^{\!\! 1/2} = \left(\frac{8 \times \frac{3}{8} \times 0.001}{(3\frac{1}{2})^2}\right)^{\!\! 1/2}$$

= 0.015 radians or about 1-degree-increment settings;

360 settings are needed to cut the profile at this distance. As the radius decreases we may use a division of more than 1 degree. Thus, the choice of the number of machine settings is a matter of judgment, as is the original error limitation of ±0.001 in. Note that this is not the total error from the theoretical cam profile since the precision of the machine tool equipment employed must be added.

10.12   MANUFACTURING ACCURACY AND CURVE CHOICE

Let us investigate[37] the relationship between the accuracy of cam fabrication and the cam curve utilized. Because of the statistical deviation from the theoretical or mathematical contour, the curve that we desire may not be the one cut. This phenomenon could occur for either a portion or the complete curve with the greatest relative difference predominantly existing with *small cams* or *large fabricating tolerances*. Dynamic effects will not be discussed since they will be shown in the next article. The information is presented primarily as a guide for the designer to realize the proximity of his curve to other curves and compare them. Reduced work may result from choosing a different curve.

In Chapter 2, we saw that all the symmetrical basic curves are similar in shape (Fig. 10.9a). Let us use the straight-line curve as the basis for measurement. Let ∂ equal the maximum difference between the straight line and any other curve, in. This occurs at approximately the same cam angle for all curves. T_0 equals the bilateral tolerance—the amount of dimensional variation from the theoretical cam curve, e.g., ±0.002 in. This total tolerance may consist of the machine tool precision, scallop height, errors produced by poor hand-workmanship in finishing operations, etc. The tolerance is determined primarily from experience. If h is the total rise of the cam follower (in.), we could show that for each curve the ratio ∂/h is a constant. Table 10.4 tabulates the results for some basic curves and also the relationships to be used for investigating the curve or range resulting from fabricating with a tolerance T_0. Let us explain the curve deviation column of Table 10.4. As an extreme condition, we see (Fig. 10.9b) that the tolerance T_0 should not exceed the maximum difference ∂. Otherwise the curve cut could have any shape at all. It would be useless to attempt to tabulate or cut the curve with high accuracy.

For greater refinement (Fig. 10.9c), it is arbitrarily suggested that the tolerance shall be less than $\partial/3$. This guarantees following the curve statistically for more than 85 percent of the range. In a similar manner any two curves may be compared by their differences, $\partial_1 - \partial_2$

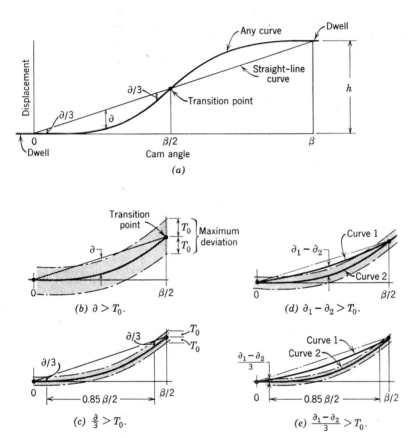

Fig. 10.9.   Maximum deviation of cam curves.

(Fig. 10.9d) or $(\partial_1 - \partial_2)/3$ (Fig. 10.9e). In this way, we may determine whether the desired curve is being cut or if the tolerance is likely to produce another curve. Note that this may not necessarily be detrimental, except in the range of infinite pulse which generally occurs at the dwell ends. No information is presented in this article about these critical ranges.

Substituting into the relationships given in Table 10.4 indicates that the tolerance discussion of this article is primarily pertinent to small cams, ramps, pre-cams, or large tolerance cams. The analysis could be elaborated to include combination and polynomial curves. Lastly, the presentation may be expanded by plotting displacement values in which Table 4.2 may be helpful.

Table 10.4   Tolerance vs. Curve

Percent Maximum Deviation from Straight-Line Curve		Curve Deviation	
Curve	∂/h	Dimensional Difference between:	Relationship
Cubic no. 2	0.10	Curve and straight line (Fig. 10.9b)	$\partial > T_0$
Simple harmonic motion	0.11		
Parabolic	0.13	Curve and straight line for 85% range (Fig. 10.9c)	$\partial/3 > T_0$
Cycloidal	0.16		
Cubic no. 1	0.20		
		Curves 1 and 2 (Fig. 10.9d)	$\partial_1 - \partial_2 > T_0$
		Curves 1 and 2 for 85% range (Fig. 10.9e)	$\dfrac{\partial_1 - \partial_2}{3} > T_0$

Example

A high-speed cam with a total rise of 0.094 in. is needed. The cam curve chosen is the cycloidal, and the best accuracy attainable with the equipment available (including removal of increment cutter scallops) is ±0.002 in. Investigate the justification of applying this curve.

Solution

First, let us compare the cycloidal curve chosen with the straight line for 85 percent range, utilizing Table 10.4 and Fig. 10.9c.

$$\partial/3 > T_0$$

$$\frac{\partial/h \times h}{3} > T_0$$

$$\frac{0.16 \times 0.094}{3} = 0.005 \text{ in.} > 0.002 \text{ in.}$$

Therefore the curve is away from the straight-line curve, for definitely more than 85 percent of the action. Next we shall compare the curve to the simple harmonic motion, using Table 10.4 and Fig. 10.9d:

$$\partial_1 - \partial_2 > T_0$$

$$0.16 \times 0.094 - 0.11 \times 0.094 > 0.002$$

$$0.005 > 0.002$$

This shows that the cycloidal and the simple harmonic motion curves are far apart for quite a range of action. Lastly, using the 85 percent range (Fig. 10.9e),

$$\frac{\partial_1 - \partial_2}{3} > T_0$$

$$\frac{0.005}{3} < 0.002$$

which indicates that the curves are apart for less than 85 percent of the range.

In conclusion, it does not appear necessary to use the cycloidal curve, and the simple harmonic motion curve would be acceptable if desired. The accuracy of the cycloidal curve is doubtful at the dwell ends, where it is of critical concern at high speeds.

10.13 DYNAMIC EFFECT OF ERRORS

With the present methods of fabrication refined theoretical investigation of the cam curve choice is limited by the accuracy of forming or cutting the contour. With gears the accuracy has been established by experience and tests, but with cams very little information exists. The determination of whether a cam profile is satisfactory is based on the acceleration curve. Since the acceleration is the second derivative of lift, the large magnification of error will appear on it. Also we know that the acceleration curve is proportional to the force acting on the system. Thus the follower system feels the acceleration disturbances, and any cam having a satisfactory acceleration curve cannot cause dynamic disturbances to the follower.

Applying the method of finite differences[39] we can establish the approximate acceleration effect of simple profile inaccuracies. We know that the acceleration of any point b midway between two other points a and c at a small angle interval $\Delta\theta$ is

$$a_b \cong \left(\frac{\omega}{\Delta\theta}\right)^2 (y_a + y_c - 2y_b) \quad \text{in./sec}^2 \qquad (6.32)'$$

where y = follower displacement, in.

σ = error or deviation from the theoretical cam profile, in.

Equation 6.32 may be rewritten for the three points (Fig. 10.10) which have deviations from the theoretical cam profile. The deviation of the follower acceleration from the theoretical or primary value

$$a_\sigma \cong \pm \left(\frac{\omega}{\Delta\theta}\right)^2 (\sigma_a + \sigma_c - 2\sigma_b) \tag{10.5}$$

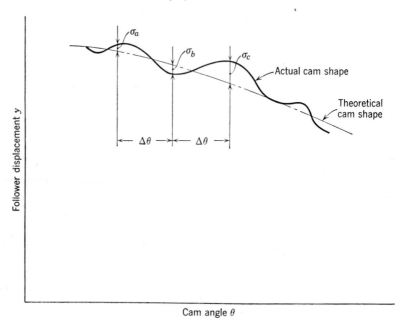

Fig. 10.10.   Displacement diagram showing cam profile deviation.

Two general cases may be approximated. The first is with the errors at each point the same value ($\sigma_a = \sigma_b = \sigma_c$) and gives the acceleration, substituting in eq. 10.5

$$a_\sigma \cong \pm 4 \left(\frac{\omega}{\Delta\theta}\right)^2 \tag{10.6}$$

The second error type is a single "dip" at point b with $\sigma_a = \sigma_c = 0$ and yields acceleration

$$a_\sigma \cong \pm 2 \left(\frac{\omega}{\Delta\theta}\right)^2 \tag{10.7}$$

Equations 10.5, 10.6, and 10.7 provide reasonable results having the *actual* cam profile deviation from the theoretical cam shape.

As a check on these relationships the author made two dynamic tests of non-periodic, single, small errors to determine the comparison between the actual and theoretical acceleration curves. The procedure used to examine the dynamic behavior of cams includes a velocity-type pickup mounted to sense the rise and fall of the cam as it rotates at a constant speed. The electric impulses from the velocity pickup were transmitted to an oscilloscope which translated the impulse to the velocity diagram. By means of integrating and differentiating circuits it was possible to convert the velocity traces into lift and acceleration traces.[24]

In Fig. 10.11 we see the theoretical cam acceleration curves of a high-speed aircraft engine valve gear linkage having a design dynamic limit WG equal to 800 pounds. Superimposed on this curve (Fig. 10.11a) is the acceleration trace of a cam profile with a single smooth non-periodic error of 0.0005 in. in 0.020 in. length of cam profile. Furthermore, the error was introduced at the top of the cam rise. We see by the acceleration profile trace that this error is only slightly larger than other profile imperfections.

In Fig. 10.11b we see the acceleration trace of a cam having a more abrupt 0.002 in. error in the same profile length of 0.020 in. This magnitude of error causes an appreciable dynamic disturbed acceleration curve.

Next, let us consider the relationship between the error acceleration and the cam profile acceleration values. From the author's experience it is suggested for satisfactory performance at high speeds (Art. 9.8) that the allowable error acceleration

$$a_\sigma \leq (20 \text{ to } 50\%)a_m \qquad (10.8)$$

where a_m = maximum follower acceleration, in./sec^2. It should be remembered that both accelerations a_σ and a_m are proportional to the square of the cam speed.

Obviously the error location with respect to the theoretical cam acceleration curve may affect the performance and life of a cam-follower surface. Taking the cycloidal cam as an example, we note that the contour accuracy must be very high. An error of a few thousandths of an inch in initial and final stages of the rise will definitely affect the follower performance. The reader should realize that the information presented in this article is introductory into the effect of surface irregularities which, of course, are dependent on the size, type, shapes, and frequency of errors, surface finish, oil film thickness, modulus of elasticity of materials, natural frequency of the follower, and method of fabrication, among other factors. Investigations should be carried further.

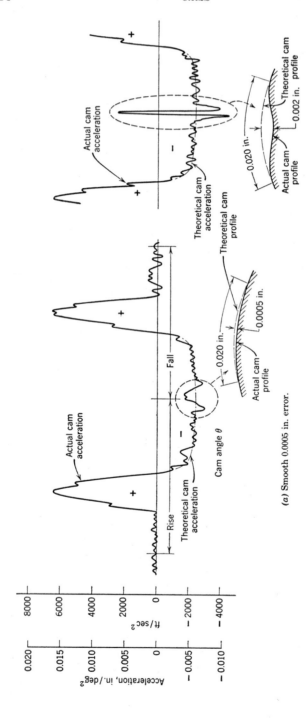

(a) Smooth 0.0005 in. error.

(b) 0.002 in. error.

Fig. 10.11.   Dynamic effect of single errors arbitrarily located on a 14-in.-diameter four-lobe high-speed aircraft valve gear cam.   Linkage weighs 4.25 lb.   Maximum lift = 0.600 in.; maximum follower velocity = 11.9 ft/sec.   Designed cam speed = 340 rpm.   (Note 0.020-in. arc is ¾ degree.)

Example

By the mathematical relationships developed, verify the test results of Fig. 10.11.

Solution

In Fig. 10.11a the 0.0005 in. error has a measured acceleration $a_\sigma = 8400$ in./sec^2. By eq. 10.7

$$a_\sigma \cong \pm 2\sigma \left(\frac{\omega}{\Delta\theta}\right)^2$$

$$\cong \pm 2(0.0005) \left(\frac{363 \times 2\pi}{60 \times \frac{3}{4} \times \pi/180}\right)^2 = 8400 \text{ in./sec}^2$$

This shows good correlation between test and calculated values. In Fig. 10.11b the 0.002 in. error has a measured acceleration $a_\sigma = 45,000$ in./sec^2. Again substituting in eq. 10.6

$$a_\sigma \cong \pm 2(0.002) \left(\frac{363 \times 2\pi}{60 \times \frac{3}{4} \times \pi/180}\right)^2 = 33,800 \text{ in./sec}^2$$

which shows an acceptable discrepancy between the test and calculated values.

Utilizing eq. 10.8 we could see that the 0.002-in. error gives a dynamically excessive effect whereas the 0.0005-in. error provides reasonable acceleration values.

10.14 SURFACE FINISH

Finishing of cams is often the most critical of the fabricating operations. Many cams require no operation after the jig borer, milling, or grinding machine cutting stages. Others may need hand filing or stoning of scallops or flats, especially for the master or leader cams. In general, any hand operation is subject to error and is completely dependent upon the operator's skill and experience. Hand lapping or finishing is not suggested for high-speed cams, unless the ability of the craftsman is assured. Otherwise, the deviation from the desired cam shape, dynamically speaking, may actually increase instead of being eliminated.

The next point of consideration in cam design is, "How smooth must a surface be for the ultimate in material life and follower performance?" First, let us state that very little information is available on this subject. Obviously surface wear and follower performance are related. It is known that rough surfaces wear more than smooth parts, but the degree of smoothness may vary. With rough surfaces, the microstresses will be high, since the contacting surface asperity area is low. Thus, failure generally originates at these high points. It is stated[12] that polishing of the automotive sliding follower to a maximum of 4 microinches is used whereas cam finishes as high as 50 microinches are acceptable. Roller followers have been finished as smooth as 4 microinches. Reduced roughness of surface-ground cams has in some instances appreciably reduced wear. If a surface needs improvement, a shear polish, lap, and controlled run-in period are suggested. However, polishing is another step that is limited by cost. Some manufacturers use a scratch pattern in the surface to trap the oil resulting in a longer life for the cam and the roller follower. Sometimes a chrome-plated surface over a polished one will further increase wear life. This is especially true when the lubrication of the surfaces in contact (under high sliding velocities) is doubtful.

10.15 CHECKING CONTOUR ACCURACY

Rarely does the actual cam manufactured agree with the theoretical mathematical profile desired. When the differences exceed a reasonable minimum value (determined by the cam function and application) we speak of them as errors. These errors require that highly accurate inspection equipment be employed to check the profile. Cam contour inspection methods may be divided into three types: *template, micrometer (indicator) measurement,* and *electronic.*

The *template method,* as the name implies, utilizes an accurate template to be compared with the actual cam. This method is generally not suggested because the accuracy is limited and it cannot be applied to production inspection. However, magnifying optical comparators using these principles have given surprisingly precise measurements.

The *micrometer method* has a suitable fixture recording the cam contour lift on a dial indicator. This approach of measuring a contour to be checked against a master cam is time consuming and impractical for production tests. Checking the contour errors after measurement requires:

(a) A plot of the successive overlapping portions of the curve on a greatly enlarged scale in which each small division of the graph paper represents a 0.0001-in. lift. Any value that is in error will noticeably differ from the smooth curve through the neighboring points.

(b) A quicker more accurate method of checking is to take the first and second differences of tabulated lifts. A mistake in one or more of the lift values will cause large fluctuations in the differences. However, if the change is smooth and exhibits no peculiarities, the lift figures are probably correct.

(c) A third approach is to determine the acceleration effect of the error by utilizing the method of finite differences (Art. 6.16).

The *electronic method*, previously described in Art. 10.13, uses an electrical circuit with an oscillograph which records the acceleration curves of a cam. This equipment is rapid, accurate and more reliable than the others and is especially valuable for investigating high-speed machinery cams. Appraisal of the acceleration curve will indicate the follower performance in contact with the *actual* cam. In this way, excellent results are obtained since the true dynamics of the cam profile are indicated. However, it should not be construed as replacing all micrometer measurements in production inspection. A spot check of the cam size and shape may be necessary since the cam may have incorrect proportions and provide an acceptable acceleration curve.

10.16 CONCLUSIONS

Practically all machined surfaces are generated and functionally operated under the influence of impact and impulsive forces of variable intensities and frequencies. These dynamic forces which are of indeterminable magnitude affect the cam profile, accuracy, surface quality, and life. Moreover, the magnitude of these factors and the performance of the cam follower systems are greatly influenced by the fabrication of the cam profile. The interrelated factors of accuracy, cost, and time are pertinent to the study. Poor manufacturing techniques can seriously impede the functional ability of the mechanism. Under most methods of fabrication, the actual and the theoretical cams rarely agree. Accuracies as close as ±0.0003 inch may be necessary in high-speed machinery and computer mechanisms.

References

1. H. Hertz, *Gesammelte Werke*, Vol. 1., J. A. Barth, Leipzig, 1895.
2. M. C. Turkish, *Valve Gear Design*, First Ed., Eaton Mfg. Co., Detroit, Mich., p. 100, 1946.
3. H. R. Thomas and U. A. Hoersch, "Stresses Due to the Pressure of One Elastic Solid on Another," *Eng. Expt. Sta., Univ. Illinois Bull. 212* (1930).
4. F. B. Seely and J. O. Smith, *Advanced Mechanics of Materials*, Second Ed., John Wiley & Sons, New York, 1952, p. 342.

5. J. O. Almen, "Surface Deterioration of Gear Teeth," *ASM Conference on Mechanical Wear, M.I.T.*, p. 229, 1950.

6. E. Buckingham, *Analytical Mechanics of Gears*, First Ed., McGraw-Hill Book Co., New York, 1949, p. 502.

7. E. Buckingham and G. J. Talbourdet, "Recent Roll Tests on Endurance Limits of Materials," *ASM Conference on Mechanical Wear, M.I.T.*, p. 289, 1950.

8. G. J. Talbourdet, "Laminated Phenolics on Cams," *Prod. Eng. 10*, p. 369 (Sept. 1939).

9. G. J. Talbourdet, "A Progress Report on the Surface Endurance Limits of Engineering Materials," *ASME, ASLE, 54-LUB-14*, Baltimore, Md. (Oct. 18, 1954).

10. E. Buckingham, "Qualitative Analysis of Wear," *Mech. Eng., 67*, p. 576 (Aug. 1937).

11. S. Way, "Pitting Due to Rolling Contact," *Trans. ASME 57*, p. 249 (1935).

12. V. Ayres, "Question and Answer Discussion on Valve Gear Problems," *Eaton Forum, Eaton Mfg. Co., Detroit, Mich. XIV*, 6 (Dec. 1953).

13. T. Barish, "Ball Bearings Used as Cam Rollers," *Prod. Eng. 8*, p. 2 (Jan. 1937).

14. J. T. Richards, "What Beryllium Copper Offers the Designer," *Mach. Des. 21*, p. 117 (Aug. 1949).

15. J. Holt, "Don't Underrate Nylon for Mechanical Components," *Prod. Eng. 24*, p. 203 (June 1953).

16. J. B. Bidwell and P. Vermaire, "Lifter and Lubricants," *SAE Nat. Pass. Car Body & Mtls. Meeting*, Detroit, Mich., March 2, 1954.

17. T. W. Havely, C. A. Phalen, and D. G. Bunnell, "Influence of Lubricant and Material Variables on Cam and Tappet Surface Distress," *SAE Pass. Car Body & Mtls. Meeting*, Detroit, Mich., March 2, 1954.

18. F. J. Lennon, Jr., "Designing for Wear Resistance with Cemented Carbides," *Mach. Des. 26*, p. 176 (Feb. 1954).

19. Catalogue No. 39, *Meehanite Cams, Camshafts and Crankshafts*, Meehanite Metal Corp., New Rochelle, N. Y.

20. E. Buckingham, "Surface Fatigue of Plastic Materials," *Trans. ASME 66*, p. 297 (May 1944).

21. R. Davies, "Hydrodynamic Lubrication of a Cam and a Cam Follower," *ASLE, ASME, Lub. Conf., 54-LUB-14*, Baltimore, Md. (Oct. 18, 1954).

22. W. Pease, "An Automatic Machine Tool," *Scientific American 187*, p. 101 (Sept. 1952).

23. Catalogue, *Automation of Machine Tools*, Turchan Follower Machine Co., Detroit, Mich.

24. M. C. Turkish, "Inspection of Cam Contours by the Electronic Method," *Eaton Forum, Eaton Mfg. Co., Detroit, Mich. XI*, 1 (March 1950).

25. W. M. Stocker, "Long Helical Cams Cut on Gear Shaper," *Am. Machinist 93*, p. 100 (Oct. 20, 1949).

26. *Practical Treatise on Milling and Milling Machines*, Brown and Sharpe Mfg. Co., Providence, R. I., 1952.

27. H. A. Fromelt, "Methods of Cam Milling," *Am. Machinist 86*, (1) "Use of Rotary Table," p. 321 (April 16, 1942); (2) "Use of the Dividing Head," p. 387 (April 30, 1942); (3) "Use of Dummy Templates," p. 440 (May 16, 1942); (4) "Use of Rotary Head Milling Machine," p. 502 (May 28, 1942);

(5) "Use of the Cam Slide," p. 652 (June 25, 1942); (6) "Use of Special Cam Slides," p. 709 (July 9, 1942).

28. J. E. Hyler, "Cams, Their Production and Application," *Mach. and Tool Blue Book 5*, p. 131, Jan. 1948; p. 143, Feb. 1948; p. 199, Mar. 1948; p. 179, April 1948; and p. 155, May 1948.

29. W. Jellig, "Precision Machines Assure Cam Accuracy," *Iron Age 173*, p. 140 (April 15, 1954).

30. R. F. V. Stanton, "Making Cams for Automatic Cigar and Cigarette Machinery," *Machinery, N. Y. 57*, p. 141 (Dec. 1950).

31. C. J. Green, *Cam and Shape Grinding*, Norton Company, Worcester, Mass., 1948.

32. P. Stoner, "Grinding Accurate Cam Surfaces," *Machinery, N. Y. 51*, p. 163 (Sept. 1944).

33. Anon., "Entire Cam Contour Checked in Forty Seconds by Optical Gaging," *Am. Machinist 97*, p. 140 (Aug. 31, 1953).

34. *A Treatise on Milling and Milling Machines*, Cincinnati Mill. Mach. Co., Cincinnati, Ohio, Third Ed., p. 502, 1951.

35. F. W. Hale, "Milling Original Contoured Shapes with Interpolating Tracer Control," *Tool Engineer 32*, p. 46 (June 1954).

36. *Cam Catalogue No. 55C*, Eisler Engineering Co., Newark, N. J.

37. K. Brunell, unpublished report of July 1, 1955.

38. M. Morgan, "Punched Tape Control of Cam-Milling Machines," *Elect. Mfg. 56*, p. 114 (Oct. 1955).

39. R. C. Johnson, "Method of Finite Differences for Cam Design," *Mach. Des. 27*, p. 195 (Nov. 1955).

CHAPTER ELEVEN

Special Cams and Applications

11.1

In this chapter we shall discuss some special cams and their applications to various kinds of machinery. Many diversified examples are presented with no attempt to cover every possible machinery application. The mathematical theory is kept to a minimum to emphasize the basic practical understanding of the subject. In some instances, comparisons are made between these novel cams and other mechanisms. Excellent compilations of practical cam mechanisms are given by Jones[1] and Grodzinski.[2]

11.2 QUICK ACTION CAMS

Sometimes machines require cam actions to provide a quick follower movement. In Art. 3.13, we have indicated that a roller follower on the conventional radial cam cannot have an instantaneous return even if the cam had a sharp corner. The path of the roller follower was an arc of a circle of radius equal to the roller radius. Now, we shall show two practical cams that do not have this shortcoming. In Fig. 11.1a, we have lugs C and C', which are fixed to the cam-shaft. The cam is free to turn (float) on the cam-shaft, limited by lug C and the adjusting screw. With the cam rotating clockwise, lug C drives the cam through the cam lug B. At the position shown the rollers will drop off the edge of the cam which is accelerated clockwise until its cam lug B hits the adjusting screw of lug C'. The action is quick but obviously is limited by the inertia of the cam mass itself and by the amount of noise that one is willing to tolerate when the follower falls completely. Figure 11.1b shows the application of two integral cams and two followers, providing much faster action. The roller follower rides on cam no. 1. At the position shown, the flat-faced follower starts on cam no. 2. Con-

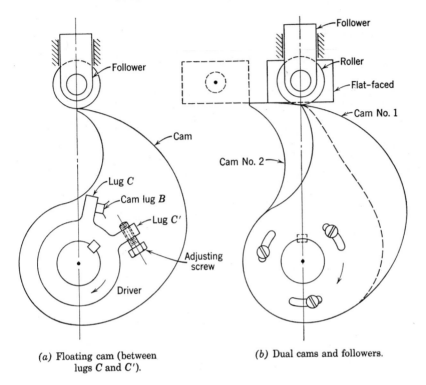

(a) Floating cam (between lugs C and C').

(b) Dual cams and followers.

Fig. 11.1.   Quick-acting cams.

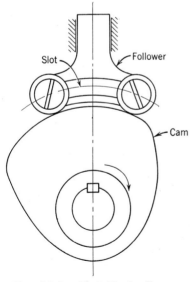

Fig. 11.2.   Variable-dwell cam.

tinued clockwise rotation of the cam will move the flat-faced follower to the edge of the cam (dotted position), where it drops off the edge. Slots shown may be employed for timing adjustment.

11.3   VARIABLE-DWELL CAM

In some special machinery, it is necessary to vary the dwell period of the cam. This may be accomplished by adjusting and fixing the distance between two rollers in the slot (Fig. 11.2). Note that changing one dwell period (rise) effects a change in the total action in which two radial cams would provide greater control.

11.4 CAM TO CONVERT LINEAR TO ROTARY MOTION

Let us now analyze a novel inverse indexing cam design utilizing a continuous trilevel cam track. In this design (Fig. 11.3), a stationary cam pin rides in the cam track to convert horizontal linear motion into rotary motion. The trilevel track guarantees the proper indexing directions without backtracking. This mechanism has been

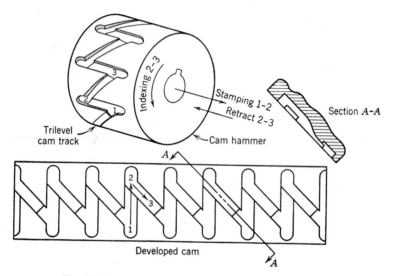

Fig. 11.3.   Cam to convert linear to rotary motion.

applied to index the stays on a stamping machine. The stationary pin (not shown) is in position no. 1 with the cam hammer assembly fully retracted. At position no. 2, the hammer has completed its stamping blow. As the hammer assembly retracks, the cam track follows the stationary pin to position no. 3 to impart rotary motion to the inverse cam.

11.5 CAM TO CONVERT ROTARY TO LINEAR MOTION

Six steel balls that cause an inverse-face cam to assume an up-and-down motion result in a vibratory motion of a shaft attached to the cam (Fig. 11.4). This reciprocating movement of the shaft has been applied in the form of a high-frequency shock to the drill core of the rotary hammer. The total shaft output was 6000 blows per min. at 1000 rpm. Contoured and convex-shaped, the grooved face of the cam contacts the exposed portion of the balls; the rest of the balls are housed in recesses of the ball seat, which at the same time acts as a spacer for the balls. Heat-treated Nitralloy is utilized to give required hardness and to minimize wear.

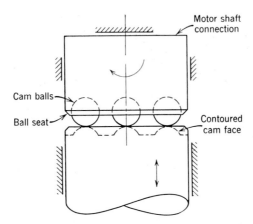

Fig. 11.4. Cam to convert rotary to linear motion.

11.6 TWO-REVOLUTIONS-PER-CYCLE CAMS

There are two types of cams that can fulfill the requirements of two revolutions of the cam for one complete movement of the fol-

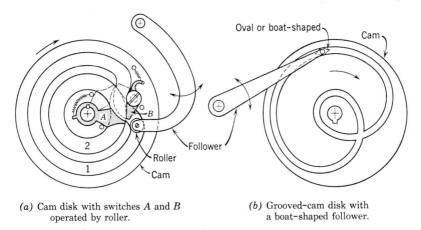

(a) Cam disk with switches A and B operated by roller.

(b) Grooved-cam disk with a boat-shaped follower.

Fig. 11.5. Two-revolutions-per-cycle cams.

lower. These cams will provide full lift of the follower with the cam rotation of more than 360 degrees. The mechanism shown in Fig. 11.5a utilizes a double-groove cam with an oscillating roller follower.

A translating follower may also be used. This cam has movable doors or switches A and B directing the follower alternately in each groove. The grooves may be designed so that we may have follower movement or dwell as required. At the instant shown door B is ready to guide the roller follower from slot **1** to slot **2**. The other door positions are shown dotted.

In Fig. 11.5b, we see a cam mechanism that fulfills the same requirements without the use of movable doors. As before, a groove cam is employed. However, the follower is made oval or boat-shaped in order to transverse the small radii of curvature and high-pressure angles that exist in this type of cam. Although a radial cam and an oscillating follower are shown, cylindrical cams or translating followers may also be utilized.

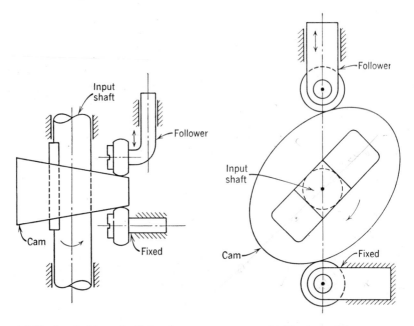

(a) Floating double-end cylindrical cam.   (b) Floating radial cam.

Fig. 11.6.   Increased stroke cams.

11.7   INCREASED STROKE CAMS

In Fig. 11.6, we see examples of cams giving an increased stroke without increasing the pressure angle. Both examples, one a radial cam and the other a cylindrical cam, are kinematically the same design. The mechanism shown in Fig. 11.6a has the input shaft parallel to the follower movement whereas Fig. 11.6b has the input shaft perpendicular to the follower movement. In both examples, the cam slides on the input shaft and is in contact with a fixed roller upon which the complete mechanism rides. Thus the total movement of the follower is the sum of the cam displacement on the fixed roller plus the follower displacement relative to the cam.

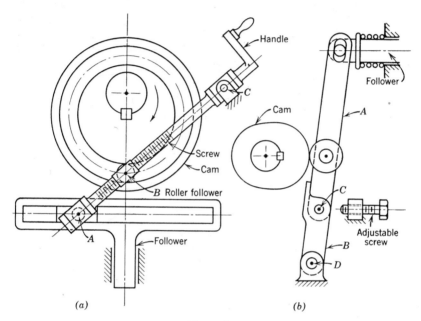

(a)

(b)

Fig. 11.7. Adjustable stroke cams—screw type.

11.8 ADJUSTABLE STROKE CAMS—SCREW TYPE

In Fig. 11.7, we see two possibilities of adjustable stroke cams in which the total follower movement can be changed while the machine is running. In Fig. 11.7a, we have a screw moving on its pivot at C, and the roller follower B driving the follower through point A. The stroke adjustment is made by turning the screw handle, which changes the critical distance AB.

Figure 11.7b shows two links A and B that are connected at point C. These links either are in line pivoting about point D or are pivoting about C contacting the fixed adjustable screw. The follower stroke variation is made with the adjustable screw. The stroke depends on the relative amounts that lever A pivots about either points C or D. Therefore, minimum stroke will occur with the adjustable screw to the right. In this instance, the arms AB, if never in contact with the screw and the lever, will always pivot about D.

Similarly, the maximum stroke will occur when the adjustable screw
is at the left so that the arms will pivot about points C and D. Thus
infinite stroke adjustment is possible within these two ranges.

11.9 CONTROLLED TRANSLATING CYCLE CAMS

In Fig. 11.8, we have a special mechanism superimposing the
motion of two cams, A and B. As an example, cam A is the main
drive cam locating the follower position in steps, and cam B is the

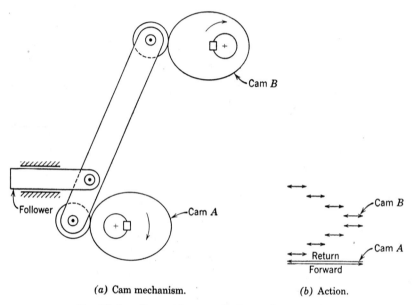

(a) Cam mechanism.   (b) Action.

Fig. 11.8.   **Cam with controlled translating cycle.**

higher-speed cam giving the follower its back-and-forth cycle for
each position. In Fig. 11.8b, we see the motion in which the cam B
turns at least 8 times as fast as cam A. We can control the translating motion cycle by varying the relative speed ratio of cams
A and B. Any combinations of motion are possible.

11.10 CIRCULAR ARC CAMS—CONSTANT-BREADTH FOLLOWER

Let us now consider a positive-drive cam composed of circular
arcs bearing against the sides of a follower track. For shapes other
than a circle, the reader is referred to Shaw.[4] If the follower completely encloses the cam, its motion will be a polygon in the shape
of the follower (Figs. 11.9a and b). Usually the follower has more

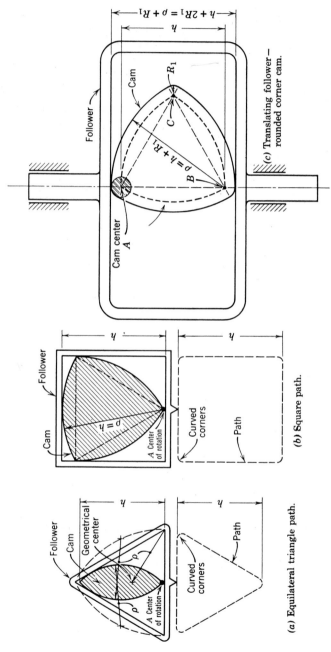

(a) Equilateral triangle path.

(b) Square path.

(c) Translating follower – rounded corner cam.

Fig. 11.9. Circular arc cams—constant-breadth follower.

sides than the cam. In Fig. 11.9, we see all the cams rotating about their centers, A, with the follower constrained to give a total rise h. In Fig. 11.9a, we have a two-sided cam and an equilateral triangle follower. The radius of cam curvature ρ equals the distance from the center to the corner of the triangle. The path of the follower is only approximately an equilateral triangle, since it has small rounded corners. In Fig. 11.9b, the cam is constructed on an equilateral triangle having a circular arc radius equal to the width of the square follower h. This follower takes a square path, again with small-radius corners. Note that the radius h is equal to the total displacement of the follower. Obviously, in a similar manner any number of sides may be used. It should be observed that a practical difficulty exists in that the centerline of the shaft A is on the cam edge, which does not always provide enough space for the shaft. This shortcoming may be alleviated by the design shown in Fig. 11.9c or by employing an unsymmetrical cam having a small radius at the cam center.

On the other hand, if the outer frame is fixed, the cam would be free to travel constrained by the frame track. Thus, any point on the cam would describe a similar geometric path, as shown in Figs. 11.9a and b. The size of the path depends on the distance from the cam center of rotation to the point on the cam under consideration. This principle has been applied in the drilling of holes of a polygon of any shape.[5]

Let us analyze a triangular circular arc cam enclosed by a translating follower on two sides only (Fig. 11.9c). This example will serve as a basis for understanding the action of all circular arc cams. It may be mentioned that it has been successfully used for sewing machines, film movement, fuel pumps, and other mechanisms. Its application is in silent, high speed, lightly loaded, small power mechanisms. Some of these have been in excellent condition after over twenty years of operation. For further information, see the discussions by Richards.[6, 7]

The design shown in Fig. 11.9c is more practical than the previous kinds because space has been provided for the shaft. Furthermore the cam has a radius R_1 in each corner. This radius has the additional advantage of increasing the wear life over the previous cams having sharp corners. Note that the actual displacement h is unaffected by the size of this radius since the basic cam is contour ABC. The sides of the cam are shown at a radius $\rho = h + R_1$, and the breadth of the follower is $h + 2R_1$.

Using this triangular circular arc cam, we shall show that the displacement is made up of parts of the simple harmonic motion curve. All action with circular arcs is similar. Let θ equal the angle of cam rotation from some original position (Figs. 11.9c and 11.10). If we consider the cycle in which θ varies from 0 to 60

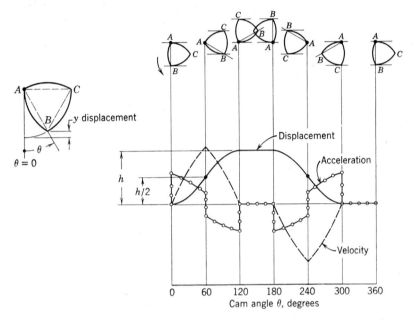

Fig. 11.10.   Triangular circular arc cam characteristics—translating follower.

degrees (the positive acceleration period of the motion), the displacement

$$y = h\,(1 - \cos\theta)\quad \text{in.}\tag{11.1}$$

This is simple harmonic motion for these 60 degrees to a total lift equal to $h/2$.

The velocity and acceleration by differentiation

$$v = \omega h \sin\theta\quad \text{ips}\tag{11.2}$$

$$a = \omega^2 h \cos\theta\quad \text{in./sec}^2\tag{11.3}$$

The minus acceleration period is similar for the next rise portion $h/2$, the total lift occurring in 120 degrees of cam rotation. A dwell of 60 degrees of cam rotation then follows. In Fig. 11.10, we plot the curves of the complete action in which it is seen that the acceleration curve is discontinuous.

11.11 SWASH PLATE CAM

The swash plate cam (Fig. 11.11) is an end cam having a flat plane disk fixed at an acute angle to a rotating shaft. This cam has been used for a multipiston pump whose piston rods ride on the swash surface. If a knife edge or point contact were employed, the follower

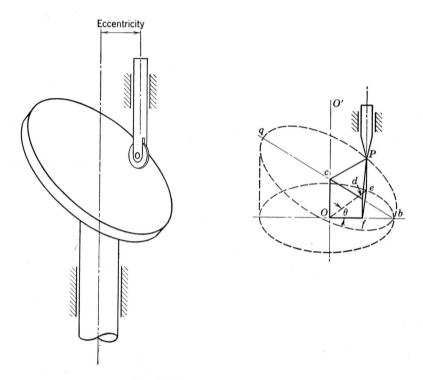

Fig. 11.11.   Swash plate cam.

would move with simple harmonic motion. However, with a more practical crowned roller follower, the follower has very nearly simple harmonic motion, if the roller diameter is kept small. Swash plate cams are feasible only under light loads, since the eccentricity shown causes poor bearing operation due to excessive deflections of the cam and its shaft.

Let us now verify the stated follower motion. If the axis of the shaft be $O\text{-}O'$ with the point P of the follower in contact with the cam, we see that, as the cam rotates an angle θ from Ob to Oe, the follower point rises a distance eP. The swash plate traces an elliptical path bPq. The rise by proportion and similar triangles is

$$eP = df = Oc\frac{bf}{bO} = Oc\left(\frac{bO - Of}{bO}\right)$$

which solves to

$$eP = Oc(1 - \cos\theta) \qquad (11.4)$$

This is harmonic motion.

11.12 COMPUTER CAMS AND LINKAGES

In Chapter 5, we discussed the use of cams and linkages in computer mechanisms. Sometimes the linkages employed as computer elements alone cannot be completely analyzed and cams are applied as

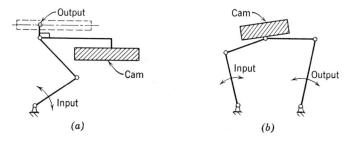

Fig. 11.12. Computer cam and linkages.

corrective devices for the mechanism (Fig. 11.12).[20] In this manner, high accuracy can be maintained, since the range of cam operation is small as compared to the total movement of the mechanism.

11.13 VARIABLE-ANGULAR-VELOCITY CAM

In all other cam mechanisms heretofore discussed in this book we have had the cam rotating at a constant speed. In practice the cam is generally geared or coupled to the constant-speed prime mover, such as an engine or an electrical motor. However, particular design requirements may necessitate a variable-speed cam to drive the follower (Fig. 11.13). The analysis of variable-speed cam mechanisms takes us into an unknown field. No data or information is available for pressure angle, forces, surface life, kinematic action, etc. Let us use as an example an oscillating variable-speed cam driven from a constant-speed crank. Combining the cam with a quick-action mechanism gives rapid follower motion without any increase in pressure angle and side thrust forces. A smaller cam, smaller pressure angle, and a compact overall mechanism result. It should be realized that this is only one example in a broad field to indicate the potentialities of variable-speed cams.

Example

Design a cam driven by a Whitworth quick-return mechanism to move a translating roller follower 1 in. with simple harmonic motion. The driver arm turns 90 degrees at a constant speed, with the cam

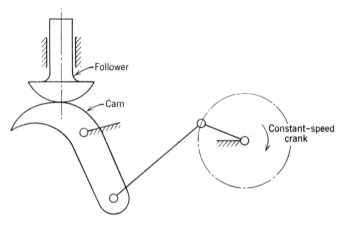

Fig. 11.13.   Variable-angular-velocity cam.

action (follower rise and fall) occurring in 180 degrees of cam rotation.   The cam pressure angle shall be limited to about 22 degrees (Fig. 11.14).

Solution

(a) Assume a driver arm 2 in. long; draw arc of arm motion CB with an included angle of 90 degrees.

(b) Draw the line of motion CB, giving the cam center at point D.

(c) The radius of the pitch circle from eq. 3.19

$$R_p = \frac{fh}{\beta}$$

$$= \frac{3.90 \times 1}{\pi/2} = 2\tfrac{1}{2} \text{ in.}$$

(d) This locates the transition point position.   Now divide the total displacement, h, of 1 in. into 6 rise and 6 fall SHM proportions (Art. 2.9).

(e) Divide the arc CB into 12 equal parts ($0'$, $1'$, $2'$, $3'$, etc.) since the driver arm turns at a constant speed.

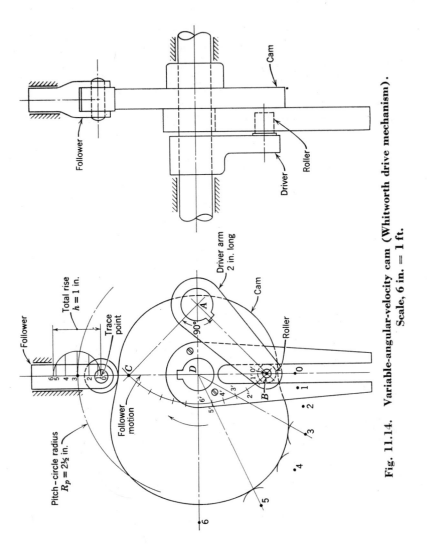

Fig. 11.14. Variable-angular-velocity cam (Whitworth drive mechanism). Scale, 6 in. = 1 ft.

(f) Draw radial lines for the cam center D to the 12 intercepts. These are the radial lines for the cam SHM displacement.

(g) Swing the SHM divisions to these radial lines, giving trace point positions 1, 2, 3, etc.

(h) Draw the roller follower arcs, and construct the cam.

(i) Construct the Whitworth mechanism.

In this manner, with a pressure angle limitation of 22 degrees, we were able to construct a smaller cam than the conventional one since the rise occurs in one-half the angle or one-half the time of the driver movement.

11.14 CAM MECHANISMS FOR SPEED CONTROL

Many fields including the cloth, wire, and paper machinery need to have the machine's speed controlled over a cycle of operation. This requirement may often be best satisfied by combining cams with gears, all driven by a constant-speed motor or engine. Accuracy, reliability, low vibration, and low noise may be achieved even when large masses

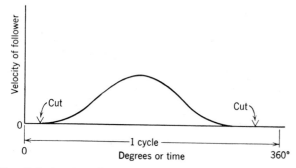

Fig. 11.15. Velocity-time diagram of paper feed mechanism (paper to be cut when stationary).

are moved at high speeds. Cams offer all the advantages of controlled design to give the best dynamic properties for the mechanism, since they allow flexibility in choice of follower action; i.e., any curve may be selected such as dwells, constant velocity, constant acceleration, and cycloidal motion. For example, in Fig. 11.15, we see a velocity curve (often called speedograph) that was a requirement for the web feeding of paper. The paper was to be cut while stationary. The problem of this intermittent paper motion may be solved by employing any of the following mechanisms.

Worm and worm gear

In this design (Fig. 11.16), the input shaft A drives the output worm gear C through its meshing worm B. Shaft A rotates and is also made to reciprocate by a fixed roller in the positive-drive cam E. Thus the output of C is made up of the rotation of shaft A plus or minus its axial movement. Any movement of output shaft C may be obtained, depending on the design of the cam E. Thus a zero velocity of the output gear is the net result of the input shaft driving in a direction opposite to that of the cam.

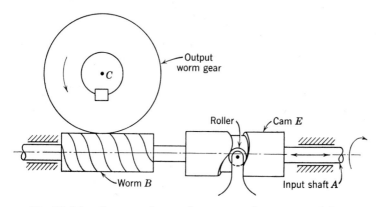

Fig. 11.16.   Cam speed-control—worm and worm gear drive.

Epicyclic gears and moving cam

Again the output speed is the sum of the input gear movement plus or minus the cam movement. In Fig. 11.17, we see a compound train with A the input gear and D the output gear. Gear D is floating, in other words not attached to gear A. Gears B and C and cam E are fixed to each other. Cam E is riding over roller F. When the unit rotates, gear D is driven by the output of gears plus or minus the movement of the arm shown. Thus we have a cam-controlled cyclical variable-speed drive.

Epicyclic gears and fixed cam

Again the output speed is the superposition of the input gear motion plus or minus the cam-driven action. Figure 11.18 shows the operation. The input A drives the arm C through pivot B. The movement of the pivot plus the movement of roller E in its fixed-cam track drives output gear D. Epicyclic action occurs in which the output gear rotation is the sum of both the rotation of gear A and the position of roller E in the fixed cam track.

Epicyclic gears and moving cam

This design is kinematically the same as the previous ones. As before, the output speed is the superposition of two actions (Fig. 11.19),

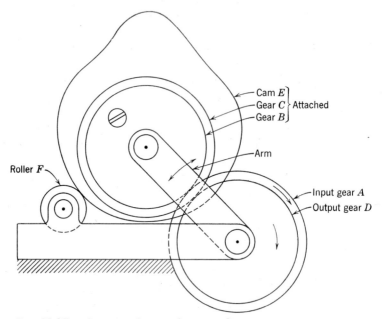

Fig. 11.17.   Cam speed-control—epicyclic gears and moving cam.

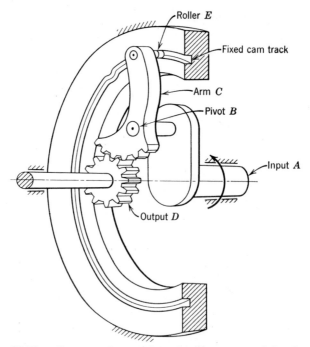

Fig. 11.18.   Cam speed-control—epicyclic gears and fixed cam.

the input gear speed and the epicyclic arm G, the latter driven by the cam. The main drive occurs through gear A into B, turning the epicyclic gears, C meshing with the output shaft on gear D. Gear C is free to rotate on its shaft. Rack F attached to the cam roller drives gear E which is attached to arm F. Thus, the output is made up of the inputs of the cam and gear A.

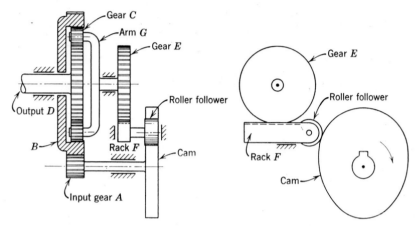

Fig. 11.19.   Cam speed-control—epicyclic gears and moving cam.

Non-circular gears

In some designs, it has been proved feasible to use non-circular gearing for a variable angular velocity ratio between the driving gear and the driven gear. Generally, non-circular gears are expensive and are utilized as a last resort. However, the previous cam and link mechanisms may become too complicated and bulky. To explain the operation, let us take an example from a printing press drive which grabs a stationary sheet of paper and transfers it at constant velocity to a constant-speed rotating cylinder. In other words, the speedograph has a constant-velocity period. Combined with non-circular gears, the cam is necessary since we have a dwell period at the beginning and at the end of the action (Fig. 11.15). This dwell period cannot be fulfilled by meshing gears alone, since the driver and driven gears must always move.

In Fig. 11.20, the follower is shown accelerating clockwise. Cam no. 1 and roller no. 1 are employed for the initial dwell range into the positive acceleration portion of the action, with cam no. 2 and roller no. 2 used for the negative acceleration portion to the dwell period. The gears mesh between these ranges of action. Note that conjugate cam mechanisms are utilized. This is an improvement over all the other designs, since backlash is kept to a minimum, resulting in smooth, shockless, and quiet action. For further discussion of this subject see Peyrbrune.[13]

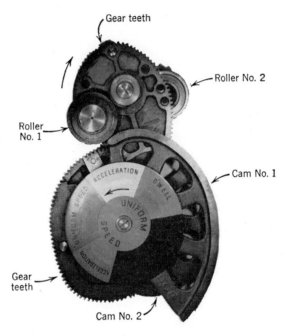

Fig. 11.20. Cam speed control—non-circular gears (printing press transfer mechanism). Courtesy Miehle Printing Press and Mfg. Co., Chicago, Ill.

11.15   COMPARISON OF INTERMITTENT-MOTION MECHANISMS

Intermittent-motion mechanisms, sometimes called indexing mechanisms or intermittors, are among the most useful devices available to machine engineers. These mechanisms have been applied to indexing dials, carriers, conveyors, feeding mechanisms, and many others. The function of these devices is to convert a continuous to an intermittent motion. The types available are the *ratchet gear, intermittent gear, Geneva* mechanism, *star wheel* mechanism, and many designs of the *cam* mechanism. In Fig. 11.21, we see the Geneva and the star wheel mechanisms; the ratchet and intermittent gear mechanisms are not shown since they are quite common. Both of these mechanisms (Fig. 11.21) index when the driver roller enters the follower slot and are held in the dwell position by the concave portion of the follower wheel.

Other mechanisms for indexing have been employed which, although ingenious, are not as practical because of the high fabrication cost and complexity of the parts. A critical survey of intermittent mechanisms was presented by Lichtwitz.[14]

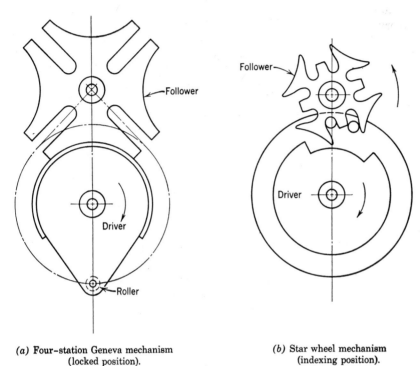

(a) Four-station Geneva mechanism (locked position).

(b) Star wheel mechanism (indexing position).

Fig. 11.21. Some intermittent-motion mechanisms.

Now, let us analyze the characteristics of the foregoing mechanisms with special concern for the acceleration curve shape. These curves have basic proportions that are specifically inherent to each mechanism. They have high peak accelerations, with either infinite acceleration or infinite pulse values (Fig. 11.22). With the ratchet gear, since pawl and wheel are not connected directly, the pawl hits the teeth of the wheel abruptly. This gives theoretical infinite acceleration followed by a modified harmonic acceleration, with the motion of the ratchet gears usually derived from a crank. Ratchet gears are used in low-speed applications where noise and accuracy of movement are not important. The intermittent gear mechanism has a similar acceleration curve and thus the same speed limitation.

Again referring to Fig. 11.22, we see that the Geneva and star wheel mechanisms are alike in that infinite pulse exists at the beginning of the action. This, as noted in previous chapters, produces vibrations, noise, and wear, which limits the speed and mass of follower. Also, with the Geneva mechanism, the driven index wheel is given high acceleration during the middle part of the movement, as the driver acts on a very small lever arm Therefore driving conditions are unfavorable, particularly with larger sizes and speeds. Special mechanisms have been combined with the Geneva to improve the acceleration curve.[2,10]

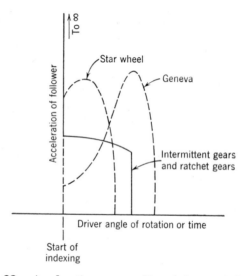

Fig. 11.22.   Acceleration curves of intermittent mechanisms.

These are complicated and expensive and thus are not feasible. The star wheel is a kindred mechanism to the Geneva, having a greater range for the rotation and dwell periods.

We shall see in the following articles that cam-driven mechanisms are best. The basic advantage is that the cam can be designed for any curve (simple harmonic, parabolic, cycloidal, or trapezoidal, etc., motions) and thus the action may be selected with complete control. By choosing the proper acceleration curve (and values) with a finite pulse at all times, we can control the inertia forces and limit the vibrations, noise, and wear under high-speed operation. Accurate and smooth indexing results. In addition, the follower may make more than one partial movement when driven by the cam.

The indexing period for these cam mechanisms may be chosen to suit, ranging from less than 90 degrees up to 360 degrees. Shorter indexing periods require larger cams: they can be used when there are no limitations in space and when the speed is not so high as to have prohibitively large inertia forces. A dwell period of about 180 degrees generally gives reasonable action and proportions.

Let us discuss the following cams and cam applications: *cylindrical, grooved concave globoidal, spider, multiple double-end,* and the *modified star wheel.* All of them are positive-drive mechanisms, with the multiple double-end cam and the star wheel cam being the best for high-speed action. These latter types fulfill the requirements of Art. 5.23, in which it was shown that opposed double conjugate rollers on the cam track give the minimum backlash and the smallest follower vibration.

11.16 CYLINDRICAL CAM FOR INTERMITTENT MOTION

The first of the cam-driven indexing mechanisms to be discussed employs a cylindrical cam.[14] This mechanism which generally has eight or more stations is applied to crossed or skewed shafts. Two of the many installations were with zipper-making equipment which had light loads at 1500 indexes per minute and with cigarette-packaging machinery allowing hopper-fed cigarettes to be inserted twenty at a time, 150 times per minute into pockets of a twelve-station turret cam, SAE 8617 carburized Rock. C-56 to 60.[15]

The driving member is a cylindrical drum having a ridge which engages the follower rollers mounted on a disk (Fig. 11.23). A part of the ridge is in a plane perpendicular to the axis of the drum. Its width is equal to the distance between the rollers, which serves to lock the driven disks. The two ends of the ridge are curved outward. The driven member is rotated by the interaction of the curved portion of the ridge with the rollers. The ridge as shown provides a positive-drive action of the driven member. However, this locking arrangement may be reversed and the basic kinematic function retained by having the two rollers guided by the curved ridge; the two straight branches of the ridge serve for locking. The reader should note the cut in the ridge to clear the next roller as the follower is indexed.

As we have previously indicated, the kinematic properties are independent of the diameter of the roller driving cam and driven disk. These dimensions may be chosen arbitrarily. However, a large cam diameter reduces the pressure angle or inclination of the ridge in the same manner as a helix angle of a screw thread. The pressure between the cam and the follower is thus lessened. In addition, the cam size is limited by the fact that as the rollers move along an arc their axes cannot point (in every position) toward the axis of the driving cam. Cutting of the grooves becomes difficult when the attempt is made to compensate for the roller position during

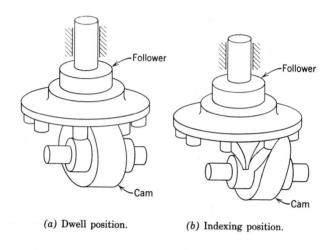

(*a*) Dwell position.   (*b*) Indexing position.

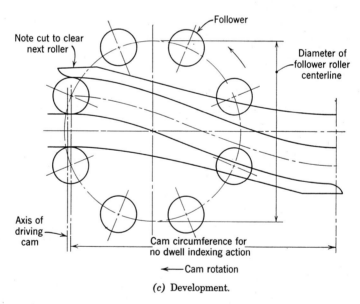

(*c*) Development.

Fig. 11.23. Cylindrical cam for intermittent motion (eight-station indexing).

action. This can be decreased by having the axis of the driving shaft midway between the extreme roller positions (Fig. 11.23).

Returning to the cam-follower action, experience has shown that quick wear due to a large amount of sliding (spalling) is evident as the rollers withdraw from large cams. Furthermore, the rollers tend to skew, owing to the end thrust. Short rollers are suggested to minimize this detrimental action. Some improvement may be made by employing conical rollers with the apex at the axis of the drum, but they are often troublesome to install.

Another hindrance of this cam mechanism is that it is often difficult to maintain locking under dwell conditions. A positive lock such as a wedged-shaped block that engages a pair of rollers during the idle period, relieving the shock load from the cam, is sometimes employed. Shock and vibration may be caused by the backlash or clearance between the rollers and their groove. In the articles that follow, we shall examine better choices for longer life, smoother performance, and possibly easier fabrication but they are often more expensive.

11.17 GROOVED CONCAVE GLOBOIDAL CAM FOR INTERMITTENT MOTION

The concave globoidal cam is often referred to as the concave barrel cam. When used for indexing, it is an improvement over the cylindrical cam. In appearance, it is similar to the cylindrical cam (Fig. 11.24). The concave globoidal cam has been successfully used in the paper industry where a 1000-lb wheel 50 in. in diameter operated 16 hr per day at 75 indexes per minute for two years without noticeable wear.

The follower has rollers that ride in grooves of the cam, giving positive-drive action during both indexing and dwell periods. These rollers remain at a constant depth as compared to the previous cylindrical cam, allowing a smaller follower wheel and thus reduced inertia. The only shortcoming of this type of groove cam follower is that the backlash between the roller and groove sides may produce noise, wear, and vibrations.

Generally, no auxiliary members, such as locating pins or wedge locks, are necessary during the dwell periods. The number of index stations on the cam-follower turret should be eight or more, to give reasonable cam and follower wheel proportions.

Experience has shown that a dwell period equal to the index period is desirable, the chordal distance being about 0.8 of the cam diameter, and the follower diameter about twice the cam diameter. An average pressure angle of about 30 degrees and a maximum pressure angle as high as 50 degrees have been successful. In Fig. 11.24b, we see a development of this cam track. For more information refer to Jacobs[19] who utilized turrets from 24 to 60 in. in diameter weighing

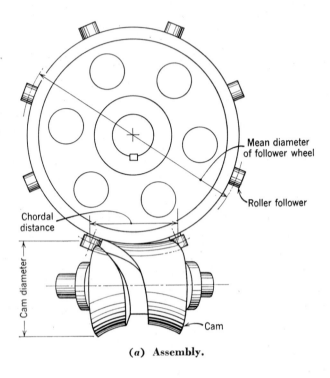

Mean diameter of follower wheel

Roller follower

Chordal distance

Cam diameter

Cam

(*a*) Assembly.

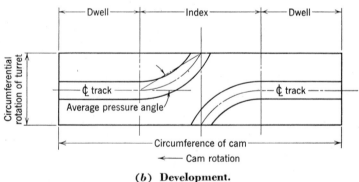

Dwell — Index — Dwell

Circumferential rotation of turret

℄ track ℄ track

Average pressure angle

Circumference of cam

← Cam rotation

(*b*) Development.

Fig. 11.24. Grooved concave globoidal cam for intermittent motion (8-station).

200 to 2000 lb. For small quantities, SAE 4140 forged steel billets with a cam track flame-hardened to Rock. C50-60 were chosen.

11.18 SPIDER CAM FOR INTERMITTENT MOTION

The spider cam[24] derives its name from the appearance of the follower. This mechanism requires a follower having four or more indexing rollers. One of its former applications was the feeding of film in motion-picture projectors.

In Fig. 11.25, we see the driver cam rotating clockwise at a constant speed driving a four-roller follower also clockwise. In Fig. 11.25a, the roller a is being moved in its slot, indexing the follower a quarter

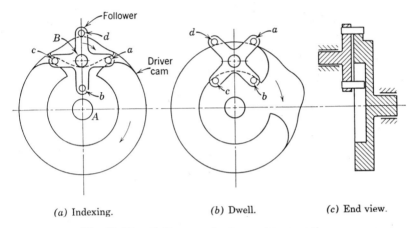

(a) Indexing.   (b) Dwell.   (c) End view.

Fig. 11.25.   Spider cam for intermittent motion.

revolution with positive-drive action. In Fig. 11.25b, we see the dwell cycle in which the rollers are positively fixed between two contours.

This cam mechanism offers more accurate control of the dwell action than the other indexing mechanisms shown. However, the limitations of backlash in the roller groove described may offer difficulty at high speeds, giving wear, noise, and vibrations in the same manner as the previous intermittent-motion cams.

The design of this mechanism may vary somewhat in that the follower pivot B may be located outside the cam ring farther from the cam center A. With this construction, the follower will rotate in a direction opposite to that of the cam.

11.19 MULTIPLE DOUBLE-END CAM FOR
INTERMITTENT MOTION

This cam having a concave globoidal shape is often referred to as the roller gear drive.[21,22]   It is available commercially.   The roller gear drive (Fig. 11.26) is similar to a worm gear driven by a worm with a varying helix angle. We see that the follower has cylindrical rollers located radially. As with other cam mechanisms, the shape of the cam rib may be made to produce any follower action with any desired acceleration curve characteristics.

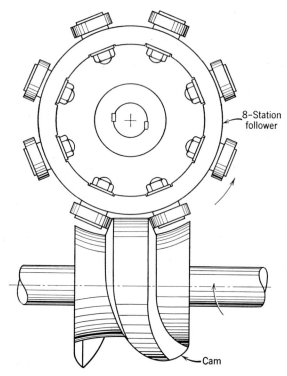

Fig. 11.26.   Multiple double-end cam for intermittent motion.

A slight change of the radial position of the rollers does not affect the performance, because they are moved along the axis parallel to the side of a tapered rib. The presence of the tapered rib permits proper roller loading, which is accomplished by shifting the driving and driven shafts closer. The use of ball bearings as rollers is advantageous, because they permit preloading against each other. They also allow building the drive with minimum backlash and with a precision of about ± 0.001 inch. The roller gear is locked at the instant at which it stops moving, so that there is no time lost for the locking operation. Thus, this cam mechanism having conjugate dual-opposed roller followers is excellent for high-speed drives, giving smooth, low-vibratory performance requiring little maintenance. It is generally expensive but often justified.

Two or more stops per cam revolution are permitted. Mechanisms that make six or more stops per revolution have the number of rollers equal to the number of stops—only one roller is passed during one revolution of the cam. These drives offer the best follower and cam proportions. If the number of stops is between two and five, then the followers should have more rollers than the number of stops. The necessity of passing more than one roller per stroke increases the size of the cam; therefore, the designer should by all means consider changing the design of the machine to permit the use of six or more stops per revolution of the gear.

This drive has been used on dial and roll feeds at speeds up to 800 pieces per minute. It has been applied to multi-color printing presses to feed paper webs intermittently at 10,000 impressions or more per hour, envelope folding machines, cap assembly machines, punch presses, and many others.

11.20 STAR WHEEL CAM FOR INTERMITTENT MOTION

The conventional star wheel mechanism is the most versatile of the intermittent mechanisms. However, as shown in Art. 11.15, the follower slot takes the path of an epicycloid which gives inherently unfavorable kinematic properties. In other words, its poor acceleration curve limits high-speed application to machinery. In the conventional star wheel mechanism (Fig. 11.21b) the drive roller is enclosed on both sides of the star wheel follower slot during indexing operation. However, a physical change in the arrangement of parts may be made to permit the use of cam-like surfaces. In this design, two rollers and two cam surfaces are employed (Fig. 11.27).

In Fig. 11.27a, we see the counterclockwise-rotating driver indexing (positive acceleration) the clockwise-rotating star wheel follower. The drive roller and the control cam are on the driving member, with the control roller and the drive cam on the driven star wheel follower. The drive roller is in contact with the drive cam, and the control roller is in contact with the control cam. Note that the drive roller contacts one side of the slot only, with clearance existing on the other side of its slot. Positive engagement, therefore, is the only function of the control roller and its contacting control cam. Fig. 11.27b shows an actual installation having the rollers on the driver and the cams on the driven member.

These preloaded cams and rollers (opposing each other) equal in performance the previous roller gear drive, providing precise action with minimum backlash. Thus high speeds and large masses may be accurately and smoothly indexed.

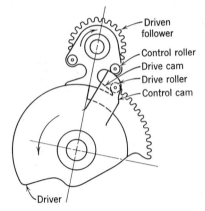

(*a*) Schematic (note cams on both).

(*b*) Unidirectional stopping mechanism applied to the drive of a printing press (note cams on driver). Courtesy Harris-Seybold Co., Chicago, Ill.

Fig. 11.27. Star wheel cams for intermittent motion.

As with all other cam mechanisms, the cam surfaces may have any form. It has been found that proper choice of acceleration curve (trapezoidal or cycloidal) will reduce the maximum acceleration value to one-half of that with the basic star wheel mechanism. A finite pulse at all times obviously reduces vibrations. This mechanism has been used in various lithographic offset and typographic presses.[23] In one of the installations, a 12-in. cylinder weighing about 1000 lb was driven with a maximum velocity of 175 rpm with one standstill period per revolution.

References

1. F. P. Jones, *Ingenious Mechanisms for Designers and Inventors,* Industrial Press, First Ed., Vol. 1, p. 1, 1948; Vol. 2, p. 1, 1948; Vol. 3, p. 1, 1951.
2. P. Grodzinski, *A Practical Theory of Mechanisms,* Emmott & Co., Ltd., Manchester, England, First Ed., 1947.
3. L. Kasper, "High Lift Cam with Low Pressure Angle," *Machinery, N. Y.,* 56, p. 193 (Sept. 1949).
4. F. W. Shaw, "Dwell Cams of Uniform Diameter," *Mech. World 98,* p. 329 (Oct. 4, 1935).
5. H. A. Dudgeon, "Drilling Square Hexagonal, Triangular and Other Polygonally-sided Holes," *Mach., London 57,* p. 429 (Jan. 16, 1941).
6. W. Richards, "The Harmonic Motion Cam of Equilateral Triangular Form," *Mach., London 57,* p. 231 (Nov. 28, 1940).
7. W. Richards, "The Triangular Harmonic Motion Cam with Square-frame Follower," *Mach., London 57,* p. 393 (Jan. 9, 1941).
8. W. Lewi, "Cam Actions for Injection Molds," *Plastic Tech. 1,* p. 27 (Feb. 1955).
9. L. A. Graham, A New Approach to Production Winding, *Mach. Des. 25,* p. 153 (April 1953).
10. R. D. Rumsey, "Redesigned for Higher Speed," *Mach. Des. 23,* p. 123 (April 1951).
11. Anon., "Cam Oscillated Gear Segment Drives Intermittent Feed," *Des. News 9* (March 1, 1954).
12. G. J. Talbourdet, "Intermittent Mechanisms," *Mach. Des. 20,* p. 159 (Nov. 1948).
13. H. E. Peyrbrune, "Application and Design of Non-Circular Gears," *Mach. Des. 25,* p. 185 (Dec. 1953).
14. O. Lichtwitz, "Mechanisms for Intermittent Motion," *Mach. Des. 23,* p. 134 (Dec. 1951); 24, p. 127 (Jan. 1952); 24, p. 146 (Feb. 1952); 24, p. 147 (March 1952).
15. R. F. V. Stanton, "Making Cams for Automatic Cigar and Cigarette Machinery," *Machinery, N. Y. 57,* p. 141 (Dec. 1950).
16. Anon., "Part Feed and Memory System Synchronized by Single Barrel Cam," *Des. News 9* (April 15, 1954).
17. Anon., "Tapered Cam-Follower Rollers Avoid Barrel-Cam Index Stroke," *Des. News 8* (Dec. 1, 1953).

18. Anon., "Drum Cam Permits Tool Reciprocation in British Punch," *Des. News 8* (Sept. 15, 1953).
19. R. J. Jacobs, "Indexing with Concave Barrel Cams," *Mach. Des. 21*, p. 92 (Feb. 1949).
20. P. T. Nickson, "A Simplified Approach to Linkage Design," *Mach. Des. 25*, p. 197 (Dec. 1953).
21. C. N. Neklutin, "Designing Cams for Controlled Inertia and Vibration," *Mach. Des. 24*, p. 143 (June 1952).
22. Catalogue, *Roller Gear Drives*, Ferguson Machine and Tool Co., St. Louis, Mo.
23. J. E. Vandeman and J. R. Wood, "Modifying Star wheel Mechanisms," *Mach. Des. 25*, p. 255 (April 1953).
24. D. D. Demans, "Cams and Spiders for Intermittent Motion," *Prod. Eng. 1*, p. 208 (May 1930).

APPENDIX A

Miscellaneous Curves

A.1 ELLIPSE

An ellipse is a curve generated by a point moving so that the sum of the distances from the two fixed points (F_1 and F_2) called foci is a constant. The basic equation for the ellipse is

$$\frac{X^2}{a^2} + \frac{Y^2}{b^2} = 1 \quad \text{(Fig. A.1)}$$

where X = value of the curve in one direction.
Y = value of the curve in other direction.
$a = \frac{1}{2}$ major axis.
$b = \frac{1}{2}$ minor axis.

We note that the foci of the ellipse F_1 and F_2 are a distance a from the ends of the minor axis. Also the major axis equals $2a$ in length. For construction of the ellipse the reader is referred to Art. 2.8.

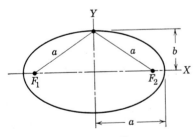

Fig. A.1. Ellipse.

A.2 PARABOLA

A parabola is a curve generated by a point moving so that its distance from a fixed point F called the focus is always equal to its distance from a fixed straight line called the directrix. In Fig. A.2, the distances X, Y, and e are shown to any point P on the curve. AB is the directrix. Since from the definition $X = e$, we have the basic

333

equation

$$Y^2 = CX$$

where C is a constant.

In addition it may be noted that with V the vertex the distance

$$GV = FV$$

Construction:

Given, in Fig. A.3, the vertex V, the axis VA, and a point P on the curve,

a) Enclose parabola in rectangle $VAPB$.

b) Divide VB and BP into the same number of equal parts, and draw lines parallel to VA.

c) Join division points on BP with vertex V.

d) The intersection of these lines gives points on the parabola VP.

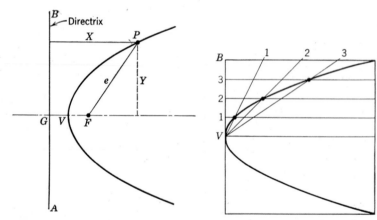

Fig. A.2.   Parabola.   Fig. A.3.   Construction of the
                                              parabola.

A.3 HYPERBOLA

A hyperbola is a curve generated by a point moving so that the difference of its distances from two fixed points called the foci is a constant.

1. General Case

The equation is:

$$\frac{X^2}{a^2} - \frac{Y^2}{b^2} = 1$$

where a and b are distances shown in Fig. A.4. Also F_1 and F_2 are the foci. Therefore, from the definition the distance $d - e = c$, a constant.

Construction:

Given foci F_1 and F_2 and distane $2a$ in Fig. A.4,

a) With F_1 and F_2 as centers, draw arcs, such as F_1P.

b) With the same centers and radius F_1P minus distance $2a$, strike arcs intersecting the other arcs giving points on the curve.

c) Repeat using new radius F_1P.

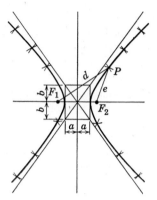

Fig. A.4. Hyperbola.

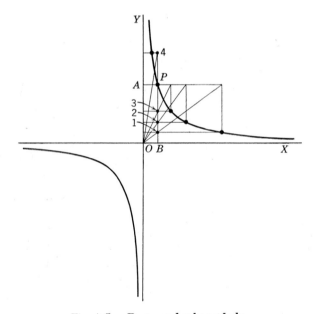

Fig. A.5. Rectangular hyperbola.

2. Rectangular

One of the most frequently employed curves in engineering is the rectangular or equilateral hyperbola (Fig. A.5).

The equation becomes

$$XY = C$$

Construction:

Given OX and OY as asymptotes of the curve and any point P on the curve,

a) Draw PA and PB.

b) Mark any points 1, 2, and 3, etc., on PB, and through these points draw lines parallel to OX and also lines through O.

c) From the intersection with line AP extended draw lines parallel to OA.

d) The intersections give the hyperbola.

A.4 LOGARITHMIC SPIRAL

This is a radial curve having a constant pressure angle. When used on a cam, it provides the smallest radial cam for a given pressure angle limitation (Fig. A.6). The polar equation is

$$r = ae^{b\theta}$$

where r = radius to any point on curve, in.

a = spiral base-circle radius, in.

$$b = \frac{1}{\tan \gamma}.$$

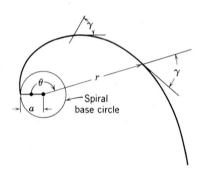

Fig. A.6. Logarithmic spiral.

$\gamma = 90 - \alpha$ = constant angle at any point between radial line and tangent to the curve, degrees.

α = pressure angle, degrees.

θ = angle between radius r and beginning point on the curve.

$e = 2.718$ = base of natural logarithms.

Construction:

Given angle γ and the base-circle radius in Fig. A.7,

a) Draw lines MP, PN, and OD.

b) Make equal division intercepts of line PM.

c) From point O, swing arc of radius BF intersecting arc of radius DF from point D.

d) This gives one point F' on the curve; other points are found in a similar manner utilizing the same centers O and D.

It may be noted that portion DQ of the curve is subject to greater error than portion DR. Caution and accuracy are suggested for this critical range.

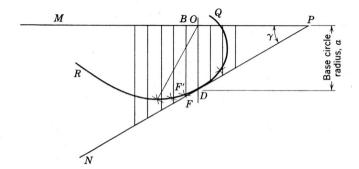

Fig. A.7.   Construction of the logarithmic spiral.

A.5   INVOLUTE OF A CIRCLE

This is a curve generated by the end of a string unwinding from an involute base circle of radius a (Fig. A.8).

The equation of the involute is

$$\theta = \left[\left(\frac{r}{a}\right)^2 - 1\right]^{1/2} - \tan^{-1}\left[\left(\frac{r}{a}\right)^2 - 1\right]^{1/2}$$

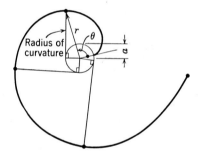

Fig. A.8.   Involute of a circle.

Construction:

Given the base circle having radius a in Fig. A.8,

 a) Divide the circle into a convenient number of parts.

 b) Draw tangents at these points.

 c) Lay off on these tangents the rectified lengths of arcs from the tangency points to the starting point.

 d) These points give the curve.

Powers 2 to 30 for
Numbers 10 to 99

In cam calculations, numbers 10 to 99 would be used as 0.10 to 0.99. Decimal points would then be located to the left of the first digit. that is, $0.10^5 = 0.00001$, $0.99^{30} = 0.73970037$, etc. Thus, values provide calculation accuracy to eight places. Where no entries appear, ciphers occur in the first eight places when the decimal point precedes the numbers.

* *Machine Design.* Courtesy Penton Publishing Co.

Table of powers n^k for $n = 10$ to 53, $k = 2$ to 16. (For the higher powers only the leading significant figures are printed; each power column carries a fixed decimal scaling, indicated by the reference row for $n = 10$. Entries are grouped in 4‑digit blocks.)

No.	2	3	4	5	6	7	8	9	10	11	12	13	14	15	16
10	0100	00 1000	0001 0000	0000 0000	0000 0100	0000 0010	0000 0000	0000 0000	0000	0000 0000	0000	0000	0000	0000	
11	121	1331	1 4641	1610	177	19	2								
12	144	1728	2 0736	2488	298	35	4								
13	169	2197	2 8561	3712	482	62	8	1							
14	196	2744	3 8416	5378	752	105	14	2							
15	225	3375	5 0625	7593	1139	170	25	3							
16	256	4096	6 5536	1 0485	1677	268	42	6	1						
17	289	4913	8 3521	1 4198	2413	410	69	11	2						
18	324	5832	10 4976	1 8895	3401	612	110	19	3						
19	361	6859	13 0321	2 4760	4704	893	169	32	6	1					
20	400	8000	16 0000	3 2000	6400	1280	256	51	10	2					
21	441	9261	19 4481	4 0841	8576	1801	378	79	16	3					
22	484	1 0648	23 4256	5 1536	1 1337	2494	548	120	26	5	1				
23	529	1 2167	27 9841	6 4363	1 4803	3404	783	180	41	9	2				
24	576	1 3824	33 1776	7 9626	1 9110	4586	1100	264	63	15	3				
25	625	1 5625	39 0625	9 7656	2 4414	6103	1525	381	95	23	5				
26	676	1 7576	45 6976	11 8813	3 0891	8031	2088	542	141	36	9	2			
27	729	1 9683	53 1441	14 3489	3 8742	1 0460	2824	762	205	55	15	4			
28	784	2 1952	61 4656	17 2103	4 8189	1 3492	3778	1057	296	82	23	6			
29	841	2 4389	70 7281	20 5111	5 9482	1 7249	5002	1450	420	122	35	10			
30	900	2 7000	81 0000	24 3000	7 2900	2 1870	6561	1968	590	177	53	15			
31	961	2 9791	92 3521	28 6291	8 8750	2 7512	8528	2643	819	254	78	24	7		
32	1024	3 2768	104 8576	33 5544	10 7374	3 4359	1 0995	3518	1125	360	115	36	11	2	1
33	1089	3 5937	118 5921	39 1353	12 9146	4 2618	1 4064	4641	1531	505	166	55	18	3	1
34	1156	3 9304	133 6336	45 4354	15 4480	5 2523	1 7857	6071	2064	701	238	81	27	5	3
35	1225	4 2875	150 0625	52 5218	18 3826	6 4339	2 2518	7881	2758	965	337	118	41	9	5
36	1296	4 6656	167 9616	60 4661	21 7678	7 8364	2 8211	1 0155	3656	1316	473	170	61	22	7
37	1369	5 0653	187 4161	69 3439	25 6572	9 4931	3 5124	1 2996	4808	1779	658	243	90	33	12
38	1444	5 4872	208 5136	79 2351	30 1093	11 4415	4 3477	1 6521	6278	2385	906	344	130	49	18
39	1521	5 9319	231 3441	90 2241	35 1874	13 7231	5 3520	2 0872	8140	3174	1238	482	188	73	28
40	1600	6 4000	256 0000	102 4000	40 9600	16 3840	6 5536	2 6214	1 0485	4194	1677	671	268	107	42
41	1681	6 8921	282 5761	115 8562	47 5010	19 4754	7 9849	3 2738	1 3422	5503	2256	925	379	155	63
42	1764	7 4088	311 1696	130 6912	54 8903	23 0539	9 6826	4 0667	1 7080	7173	3012	1265	531	223	93
43	1849	7 9507	341 8801	147 0084	63 2136	27 1818	11 6882	5 0259	2 1611	9292	3995	1718	738	317	136
44	1936	8 5184	374 8096	164 9162	72 5631	31 9277	14 0482	6 1812	2 7197	1 1966	5265	2316	1019	448	197
45	2025	9 1125	410 0625	184 5281	83 0376	37 3669	16 8151	7 5668	3 4050	1 5322	6895	3102	1396	628	282
46	2116	9 7336	447 7456	205 9629	94 7429	43 5817	20 0476	9 2219	4 2420	1 9513	8976	4129	1899	873	401
47	2209	10 3823	487 9681	229 3450	107 7921	50 6623	23 8112	11 1913	5 2599	2 4721	1 1619	5460	2566	1206	567
48	2304	11 0592	530 8416	254 8039	122 3059	58 7068	28 1792	13 5260	6 4925	3 1164	1 4958	7180	3446	1654	794
49	2401	11 7649	576 4801	282 4752	138 4128	67 8223	33 2329	16 2841	7 9792	3 9098	1 9158	9387	4599	2253	1104
50	2500	12 5000	625 0000	312 5000	156 2500	78 1250	39 0625	19 5312	9 7656	4 8828	2 4414	1 2207	6103	3051	1525
51	2601	13 2651	676 5201	345 0252	175 9628	89 7410	45 7679	23 3416	11 9042	6 0711	3 0962	1 5791	8053	4107	2094
52	2704	14 0608	731 1616	380 2040	197 7060	102 8071	53 4597	27 7990	14 4555	7 5168	3 9087	2 0325	1 0569	5496	2857
53	2809	14 8877	789 0481	418 1954	221 6436	117 4711	62 2596	32 9976	17 4887	9 2690	4 9125	2 6036	1 3799	7313	3876

339

No.	2	3	4	5	6	7	8	9	10	11	12	13	14	15	16
54	2916	15 7464	850 3056	459 1650	247 9491	133 8925	72 3019	39 0430	21 0832	11 3849	6 1478	3 3198	1 7927	9680	5227
55	3025	16 6375	915 0625	503 2843	276 8064	152 2435	83 7339	46 0536	25 3295	13 9312	7 6621	4 2141	2 3178	1 2747	7011
56	3136	17 5616	983 4496	550 7317	308 4097	172 7094	96 7173	54 1616	30 3305	16 9851	9 5116	5 3265	2 9828	1 6703	9354
57	3249	18 5193	1055 6001	601 6920	342 9644	195 4897	111 4291	63 5146	36 2033	20 6358	11 7624	6 7046	3 8216	2 1783	1 2416
58	3364	19 5112	1131 6496	656 3567	380 6869	220 7984	128 0630	74 2765	43 0804	24 9866	14 4922	8 4055	4 8751	2 8276	1 6400
59	3481	20 5379	1211 7361	714 9242	421 8053	248 8651	146 8304	86 6299	51 1116	30 1558	17 7919	10 4972	6 1933	3 6540	2 1559
60	3600	21 6000	1296 0000	777 6000	466 5600	279 9360	167 9616	100 7769	60 4661	36 2797	21 7678	13 0606	7 8364	4 7018	2 8211
61	3721	22 6981	1384 5841	844 5963	515 2037	314 2742	191 7073	116 9414	71 3342	43 5139	26 5434	16 1915	9 8768	6 0248	3 6751
62	3844	23 8328	1477 6336	916 1328	568 0023	352 1614	218 3401	135 3708	83 9299	52 0365	32 2626	20 0028	12 4017	7 6890	4 7672
63	3969	25 0047	1575 2961	992 4365	625 2350	393 8980	248 1557	156 3381	98 4930	62 0506	39 0918	24 6278	15 5155	9 7748	6 1581
64	4096	26 2144	1677 7216	1073 7418	687 1947	439 8046	281 4749	180 1439	115 2921	73 7869	47 2236	30 2231	19 3428	12 3794	7 9228
65	4225	27 4625	1785 0625	1160 2906	754 1889	490 2227	318 6448	207 1191	134 6274	87 5078	56 8800	36 9720	24 0318	15 6206	10 1534
66	4356	28 7496	1897 4736	1252 3325	826 5395	545 5160	360 0406	237 6268	156 8336	103 5102	68 3167	45 0890	29 7587	19 6407	12 9629
67	4489	30 0763	2015 1121	1350 1251	904 5838	606 0711	406 0676	272 0653	182 2837	122 1301	81 8271	54 8242	36 7322	24 6105	16 4890
68	4624	31 4432	2138 1376	1453 9335	988 6748	672 2988	457 1632	310 8710	211 3922	143 7467	97 7477	66 4684	45 1985	30 7350	20 8998
69	4761	32 8509	2266 7121	1564 0313	1079 1826	744 6360	513 7988	354 5212	244 6196	168 7875	116 4634	80 3597	55 4482	38 2592	26 3989
70	4900	34 3000	2401 0000	1680 7000	1176 4900	823 5430	576 4801	403 5360	282 4752	197 7326	138 4128	96 8890	67 8223	47 4756	33 2329
71	5041	35 7911	2541 1681	1804 2293	1281 0028	909 5120	645 7535	458 4850	325 5243	231 1222	164 0968	116 5087	82 7212	58 7320	41 6997
72	5184	37 3248	2687 3856	1934 9176	1393 1406	1003 0613	722 2041	519 9869	374 3906	269 5612	194 0840	139 7405	100 6131	72 4415	52 1578
73	5329	38 9017	2839 8241	2073 0715	1513 3422	1104 7398	806 4600	588 7158	429 7625	313 7266	229 0204	167 1849	122 0450	89 0928	65 0377
74	5476	40 5224	2998 6576	2219 0066	1642 0649	1215 1280	899 1947	665 4041	492 3990	364 3752	269 6377	199 5319	147 6536	109 2636	80 8551
75	5625	42 1875	3164 0625	2373 0468	1779 7851	1334 8388	1001 1291	750 8468	563 1351	422 3513	316 7635	237 5726	178 1794	133 6346	100 2259
76	5776	43 8976	3336 2176	2535 5253	1926 9992	1464 5194	1113 0347	845 9064	642 8888	488 5955	371 3326	282 2128	214 4817	163 0061	123 8846
77	5929	45 6533	3515 3041	2706 7841	2084 2238	1604 8523	1235 7362	951 5169	732 6680	564 1543	434 3988	334 4871	257 5550	198 3174	152 7044
78	6084	47 4552	3701 5056	2887 1743	2251 9960	1756 5568	1370 1143	1068 6892	833 5775	650 1905	507 1486	395 5758	308 5491	240 6683	187 7213
79	6241	49 3039	3895 0081	3077 0563	2430 8745	1920 3908	1517 1088	1198 5159	946 8276	747 9938	590 9151	466 8229	368 7901	291 3441	230 1619
80	6400	51 2000	4096 0000	3276 8000	2621 4400	2097 1520	1677 7216	1342 1772	1073 7418	858 9934	687 1947	549 7558	439 8046	351 8437	281 4749
81	6561	53 1441	4304 6721	3486 7844	2824 2953	2287 6792	1853 0201	1500 9463	1215 7665	984 7709	797 6644	646 1081	523 3476	423 9115	343 3683
82	6724	55 1368	4521 2176	3707 3984	3040 0667	2492 8547	2044 1408	1676 1955	1374 4803	1127 0738	924 2005	757 8444	621 4324	509 5746	417 8511
83	6889	57 1787	4745 8321	3939 0406	3269 4037	2713 6050	2252 2922	1869 4025	1551 6041	1287 8314	1068 9000	887 1870	736 3652	611 1831	507 2820
84	7056	59 2704	4978 7136	4182 1194	3512 9803	2950 9034	2478 7589	2082 1574	1749 0122	1469 1703	1234 1030	1036 6465	870 7831	731 4578	614 4245
85	7225	61 4125	5220 0625	4437 0531	3771 4951	3205 7708	2724 9052	2316 1694	1968 7440	1673 4324	1422 4175	1209 0549	1027 6966	873 5421	742 5108
86	7396	63 6056	5470 0816	4704 2701	4045 6723	3479 2782	2992 1792	2573 2741	2213 0157	1903 1935	1636 7464	1407 5999	1210 5359	1041 0609	895 3124
87	7569	65 8503	5728 9761	4984 2092	4336 2620	3772 5479	3282 1167	2855 4415	2484 2341	2161 2837	1880 3168	1635 8756	1423 2118	1238 1942	1077 2290
88	7744	68 1472	5996 9536	5277 3191	4644 0408	4086 7559	3596 3452	3164 7838	2785 0097	2450 8085	2156 7115	1897 9061	1670 1574	1469 7385	1293 3699
89	7921	70 4969	6274 2241	5584 0594	4969 8129	4423 1334	3936 5888	3503 5640	3118 1719	2775 1730	2469 9040	2198 2145	1956 4109	1741 2077	1549 6749
90	8100	72 9000	6561 0000	5904 9000	5314 4100	4782 9690	4304 6721	3874 2048	3486 7844	3138 1059	2824 2953	2541 8658	2287 6792	2058 9113	1853 0201
91	8281	75 3571	6857 4961	6240 3214	5678 6925	5167 6101	4702 5252	4279 2980	3894 1611	3543 6866	3224 7548	2934 5269	2670 4195	2430 0817	2211 3743
92	8464	77 8688	7163 9296	6590 8152	6063 5500	5578 4660	5132 1887	4721 6136	4343 8845	3996 3737	3676 6638	3382 5307	3111 9283	2862 9740	2633 9361
93	8649	80 4357	7480 5201	6956 8836	6469 9018	6017 0087	5595 8180	5204 1108	4839 8230	4501 0354	4185 9629	3892 9455	3620 4393	3367 0086	3131 3180
94	8836	83 0584	7807 4896	7339 0402	6898 6978	6484 7759	6095 6893	5729 9480	5386 1511	5062 9820	4759 2031	4473 6509	4205 2319	3952 9179	3715 7429
95	9025	85 7375	8145 0625	7737 8093	7350 9189	6983 3729	6634 2043	6302 4940	5987 3693	5688 0009	5403 6008	5133 4208	4876 7497	4632 9123	4401 2666
96	9216	88 4736	8493 4656	8153 7269	7827 5778	7514 4747	7213 8957	6925 3399	6648 3263	6382 3933	6127 0975	5882 0136	5646 7331	5420 8637	5204 0292
97	9409	91 2673	8852 9281	8587 3402	8329 7180	8079 8265	7837 4317	7602 3097	7374 2404	7153 0132	6938 4228	6730 2701	6528 3620	6332 5112	6142 5358
98	9604	94 1192	9223 6816	9039 2079	8858 4238	8681 2553	8507 6302	8337 4776	8170 7280	8007 3135	7847 1672	7690 2239	7536 4194	7385 6910	7237 9772
99	9801	97 0299	9605 9601	9509 9004	9414 8014	9320 6534	9227 4469	9135 1724	9043 8207	8953 3825	8863 8487	8775 2102	8687 4581	8600 5835	8514 5777

No.	30	29	28	27	26	25	24	23	22	21	20	19	18	17	No.
10															10
11															11
12															12
13															13
14															14
15															15
16															16
17															17
18															18
19															19
20															20
21															21
22															22
23															23
24															24
25															25
26															26
27															27
28															28
29															29
30															30
31															31
32															32
33												0000 0000	0000 0000	0000 0000	0000 0000	33
34															1	34
35															1	35
36														1	2	36
37														1	4	37
38													1	2	7	38
39													1	4	11	39
40					0000 0000						0000 0001	0000 0002	0000 0006	0000 0017	40
41											1	4	10	26	41
42										1	2	6	16	39	42
43										2	4	10	25	58	43
44									1	3	7	16	38	86	44
45							1	1	2	5	11	25	57	127	45
46								1	3	8	17	39	85	184	46
47								2	6	13	27	58	125	266	47
48						1	1	4	9	20	42	87	182	381	48
49						1	3	7	15	31	63	129	265	541	49
50					0000 0001	0000 0002	0000 0005	0000 0011	0000 0023	0000 0047	0000 0095	0000 0190	0000 0381	0000 0762	50
51				1	2	4	9	18	36	72	141	277	544	1068	51
52			1	2	4	7	15	29	56	108	208	401	772	1486	52
53		1	1	3	6	12	24	45	85	162	305	577	1088	2054	53

No.	30	29	28	27	26	25	24	23	22	21	20	19	18	17	No.
54		1	3	5	11	20	37	69	129	240	444	823	1524	2822	54
55	1	2	5	9	17	32	58	106	194	352	641	1166	2120	3856	55
56	2	4	8	15	28	50	90	161	288	515	919	1642	2933	5238	56
57	4	8	14	25	44	78	138	242	425	747	1310	2299	4034	7077	57
58	7	13	23	40	70	121	210	362	624	1076	1855	3199	5517	9512	58
59	13	22	38	65	110	186	316	536	909	1541	2612	4427	7504	1 2719	59
60	0000 0022	0000 0036	0000 0061	0000 0102	0000 0170	0000 0284	0000 0473	0000 0789	0000 1316	0000 2193	0000 3656	0000 6093	0001 0155	0001 6926	60
61	36	59	97	159	262	429	704	1155	1803	3104	5088	8341	1 3675	2 2418	61
62	59	95	153	248	400	645	1040	1678	2707	4367	7044	1 1361	1 8325	2 9556	62
63	95	151	240	382	606	962	1528	2425	3850	6111	9700	1 5398	2 4441	3 8796	63
64	153	239	374	584	913	1427	2230	3484	5444	8507	1 3292	2 0769	3 2451	5 5706	64
65	244	375	577	888	1366	2102	3235	4977	7657	1 1780	1 8124	2 7883	4 2898	6 5997	65
66	385	584	885	1341	2033	3080	4667	7071	1 0714	1 6233	2 4596	3 7267	5 6466	8 5555	66
67	605	904	1349	2013	3005	4486	6695	9993	1 4915	2 2262	3 6802	4 9593	7 4019	11 0476	67
68	944	1389	2042	3004	4418	6497	9554	1 4050	2 0663	3 0386	4 6851	6 5715	9 9640	14 2118	68
69	1463	2121	3074	4455	6457	9338	1 3563	1 9657	2 8489	4 1288	5 9838	8 6722	12 5685	18 2152	69
70	0000 2253	0000 3219	0000 4599	0000 6571	0000 9387	0001 3410	0001 9158	0002 7368	0003 8489	0005 5854	0007 9838	0011 2841	0016 1151	0023 2630	70
71	3449	4858	6842	9637	1 3574	1 9118	2 6927	3 7926	5 3417	7 5235	10 5966	14 9248	21 0208	29 6068	71
72	5247	7288	1 0123	1 4059	1 9527	2 7121	3 7668	5 2317	7 2663	10 0921	14 0168	19 4678	27 0386	37 5536	72
73	7937	1 0873	1 4801	2 0054	2 7950	3 8288	5 2450	7 1849	9 8424	13 4827	18 4695	25 3008	34 6586	47 4775	73
74	1 1938	1 6133	2 1801	2 9461	3 9812	5 3801	7 2704	9 8249	13 2769	17 9418	24 2456	32 7644	44 2762	59 8327	74
75	1 7858	2 3810	3 1747	4 2330	5 6440	7 5254	10 0339	13 3785	17 8380	23 7840	31 7121	42 2828	56 3771	75 1694	75
76	2 6570	3 4961	4 6002	6 0529	7 9644	10 4794	13 7887	18 1431	23 8725	31 4112	41 3306	54 3823	71 5557	94 1523	76
77	3 9329	5 1077	6 6334	8 6148	11 1881	14 5300	18 8702	24 5068	31 8270	41 3337	53 6802	69 7146	90 5384	117 5824	77
78	5 5921	7 4258	9 5202	12 2054	15 6480	20 0615	25 7199	32 9743	42 2747	54 1984	68 4851	89 0835	114 2096	146 4226	78
79	8 4881	10 7444	13 7543	17 2159	21 6480	27 5852	34 9180	44 2000	55 9494	72 8221	89 6482	113 4787	143 6440	181 8279	79
80	0012 3794	0015 4742	0019 3428	0024 1785	0030 2231	0037 7789	0047 2236	0059 0295	0073 7869	0092 2337	0115 2921	0144 1151	0180 1439	0225 1799	80
81	17 9701	21 1853	27 3892	33 8139	41 7455	51 5377	63 6268	78 5516	96 9773	119 7251	147 8088	182 4800	225 2839	278 1283	81
82	25 9666	31 6666	36 6178	47 6178	57 4238	70 0400	85 4146	104 1642	127 0295	154 9140	188 9196	230 3897	280 9631	342 6379	82
83	37 3544	45 0054	54 8263	65 3293	78 7100	94 8314	114 2547	137 6563	165 8509	199 8204	240 7475	290 0572	349 4665	421 0440	83
84	53 5030	63 6941	75 5379	90 2694	107 4636	127 9328	152 3010	181 3107	215 8461	256 9596	305 9043	364 1719	433 5379	516 1166	84
85	76 3075	89 7736	89 3476	124 2541	146 1813	171 9780	202 2541	238 0319	280 0376	329 4560	387 5953	455 2828	536 4640	631 1342	85
86	108	126 0245	146 5401	170 3955	198 1343	230 3887	267 8939	311 5045	362 2145	421 1797	489 7438	569 4696	662 1739	769 9697	86
87	153	176 2212	202 5531	232 8197	267 6089	307 5964	353 5591	406 3898	467 1147	536 9134	617 1419	709 3585	815 3546	937 1892	87
88	216	245 4694	285 2233	316 9425	360 9802	409 9236	465 5404	528 5687	528 1585	682 3466	775 2996	881 3953	1001 5856	1138 1655	88
89	303	340 6514	382 7543	430 0611	483 2147	542 9379	610 0425	685 4411	770 1585	865 1898	972 7665	1092 0572	1227 4715	1379 2090	89
90	0423 9115	0471 9369	0523 3476	0581 4973	0646 1081	0717 8979	0797 6644	0886 2938	0984 7709	1094 1898	1215 7665	1350 8517	1500 9463	1667 7181	90
91	590	648 0171	713 1140	783 6417	861 1448	946 3130	1039 9043	1142 7520	1255 7715	1379 9687	1516 4491	1666 4276	1831 2391	2012 3507	91
92	819	890 0934	1097 4097	1052 6193	1144 1514	1243 3603	1351 7857	1469 3323	1003 1469	1735 9786	1886 1597	2051 4134	2635 3635	2423 1257	92
93	1133	1219 0050	1310 7581	1409 4173	1515 5025	2129 5726	1752 2286	1884 1167	2025 9320	2178 4215	2901 3887	2518 6975	2708 2769	2912 7983	93
94	1562	1662 6514	1768 7543	1881 0611	2001 3552	2129 1013	2265 0014	2563 5760	2786 9985	2726 9986	3584 2866	3086 2366	2304 2304	3492 2033	94
95	2146	2259 7736	2378 3476	2503 4408	2635 2009	2773 8957	2919 8902	3073 5686	3354 3354	3405 6162	3773 8592	3360 5360	1431 3972	4181 2033	95
96	2938	3061 0171	3188 1140	3321 4161	3459 8084	3603 9671	3754 1324	3910 5546	4073 4944	4243 2233	4420 0243	4604 1920	4796 0333	4995 8680	96
97	4010	4134 0934	4097 4097	4393 6193	4529 6546	4669 7470	4814 1722	4963 0641	5116 5609	5116 8051	5437 9434	5606 1272	5779 5126	5958 2604	97
98	8431	5566 1665	7617 7581	6752 5795	9543 5913	6034 6034	8033 6157	4728 6283	7069 7069	6542 6411	6676 5021	6821 5433	6951 3533	7093 2176	98
99	7397 0037	7209 7471	1928 3476	4271 7623	4314 7700	2135 7778	7814 7856	1428 7963	3058 8016	8097 2786	8179 0693	8261 6862	8345 1376	8429 4319	99

342

INDEX

343

76